Evolution, Religion and the Unknown God

Georges Van Vrekhem

Contents

1. Darwin: The Great Amateur

The Fact of Evolution	9
Darwinism and 'Darwinism'	11
A Sea-Going Naturalist	13
Evolution in the Air	16

2. The Making of a Theory

The Smile of the Cheshire Cat	20
Darwin's Conversion	22
Gradually, Gradually	23
A Struggle for Survival	26

3. The Origin of Species

Forced to Publish	28
The Book	30
The Darwin Sect	32
What Darwin Really Said	35
What Darwin Could Not Know	38
Darwin's Legacy	41

4. Lamarck: The First Evolutionary Theorist

The Natural Sciences	47
Founding Biology	48
God or Nature	51
The Progressive Upward Climb of Life	54
Transformism	57
Caricature	60

5. *Alfred Wallace: The Other Darwin*

The Amazon and the Malay Archipelago   63
A Theory of Evolution   67
Wallace Breaks with Darwin   71
'Intelligent Design'   73
Spiritism   75
A Revival of Occultism   77
Spiritism and Science   78
Humanist – Socialist   82

6. *The Chain of Being*

Levels of Being   86
"The Serial Kingdoms of the Graded Law"   89
The Ways of the West   90
Arthur Lovejoy   91
"Man in the Middle"   93
An Idea Lives On   97

7. *Inventing 'Darwinism'*

Darwinism on the Ropes   101
A Hollow Theory?   103
A Passion for Peas   106
Cutting the Tails of Mice   108
Mutations and Saltations   111
"The Modern Synthesis"   113
Darwinism Triumphant?   117

8. *Social Darwinism*

The Copernican Principle   123
Social Darwinism   125
Practicing Social Darwinism   128
Breeding a Superman   130
"Give Me Your Tired, Your Poor ..."   133
Animal Lovers   137

9. Sociobiology

Edward Wilson – The Rise of the Genes   139
Reinterpreting the World   143
Richard Dawkins – The Triumph of the Genes   145
The Cultural Animal   149
Memes – The Gremlins of the Mind   152
A Devil's Chaplain   157

10. The Darwin Wars

Stephen Gould and Punctuated Equilibrium   161
Mass Extinctions and the Dinosaurs   166
Of Human Arrogance   167
The Darwin Wars   169

11. The Scientific Method

The Scientific Method   173
Is There a Scientific Method?   179
Parading the Paradigm   182
Biology and Physics   185

12. Of Genes, Genetics and Genomes

From Complexity to Perplexity   192
Lamarck's Come-Back   199
The "Holy Grail" of Biology   201
Epigenetics   203

13. Science and Religion

The Western God   208
Religion or Spirituality   213
Empiricism and Religion   216
The End of Science?   218
Scientists Pro and Contra   221
The Unknown God   225
A Diminished God   229

Omnipresent Reality   234

14. Intelligent Design

The Anthropic Principle   237
Intelligent Design and the Other Evolutionary Theories   240
Intelligent Design Theory   247
Irreducible Complexity   251
Problems of Intelligent Design   256

15. Intelligence that is Consciousness that is Being

West and East   259
Involution and Evolution   263
Mother Nature   268
Evolution Two-Tiered   273
Homo sapiens   277

Bibliography   283
Biographical Note   287
Index   288

1.

Darwin: The Great Amateur

It seems that if an idea is repeated often enough, then however counter-intuitive it may be, people eventually come to accept it, and to believe that they understand it.

PAUL DAVIES

The Fact of Evolution

Evolution is now generally considered a fact, except by some literalist or fundamentalist religions. "It has to be said emphatically: the theory of evolution is true and no manoeuvring will destruct its foundations, even if it is true that we do not yet understand all the mechanisms nor even all the modalities," writes Claude Allègre.[1] And another scientist, Michael Ruse, states: "By any understanding of the terms, evolution is a well-established fact. It is logically possible that evolution is not true, but it is not reasonable to believe this."[2] Evidence of the increase in the number of life-forms and their diversification has been abundantly found in the fossil record, the fact of anatomical structures common to various species, their geographical distribution, similarities during embryonic development, and DNA sequences.

In the general mind the idea of evolution is supposed to have originated with Charles Darwin, at a time that people still wore top hats, carried walking sticks and rode in horse-drawn carriages. But Darwin's idea was preceded by a lot of research and theorizing, and to a considerable extent the result of it. And although it is true that in Europe the origin of the universe, life and the human being was for many centuries attributed to a Creator "in the beginning," several cultures held that the world and everything in it had evolved.

---

1. Claude Allègre: *Dieu face à la science,* p. 140.
2. Michael Ruse: *Darwinism and its Discontents*, p. 51.

One such culture was that of the ancient Greeks, who were the first (in the West) to propose answers to the basic questions of existence in a rational, original manner. Their common view was that the history of humanity and the world was cyclic; it repeated itself again and again at huge time-intervals from chaos to cosmos to chaos. But in the fragments left us from some early Greek thinkers one can find hints of a progressive evolution. Anaxagoras of Miletus is quoted as having taught that originally humans were born from animals. Archelaus, the first Athenian philosopher and a teacher of Socrates, said that at first men were included among the animals and were afterwards separated from them. And Democritus, also a contemporary of Socrates, gave a fairly consistent picture of the evolution from primitive hunter-gatherers to agricultural civilization.

Then there is the remarkable case of Jalal ad-Din ar-Rumi (1207-1273), the great Sufi poet who, at the time of the Middle Ages in Europe but of a great cultural flowering in the Muslim world, wrote his *Spiritual Couplets*. Although orthodox Islam is creationist, we find in Rumi the following lines expressing a distinct evolutionary view:

> I died as inanimate matter and arose a plant,
> I died as a plant and rose again an animal,
> I died as an animal and arose a man.
> Why then should I fear to become less by dying?
> I shall die once again as a man
> To rise an angel perfect from head to foot![3]

However, the clearest formulation of an evolutionary vision of the universe and life on the Earth is found in the Indian scriptures. "In certain respects the old Vedantic thinkers anticipate us," wrote Sri Aurobindo, "they agree with all that is essential in our modern ideas of evolution. From one side all forms of creatures are developed; some kind of physical evolution from the animal to the human is admitted in the Aitareya [Upanishad] ... The Puranas admit the creation of animal forms before the appearance of man and in the symbol of the Ten Avatars trace the growth of our evolution from the fish through the animal, the man-animal and the developed human being to the different stages of our present incomplete evolution. But the ancient Hindu, it is clear, envisaged this progression as an enormous secular movement covering

---

3. Translation E.H. Whinfield, 1898.

more ages than we can easily count ... It is this great secular movement in cycles, perpetually self-repeating, yet perpetually progressing, which is imaged and set forth for us in the symbols of the Puranas."[4]

What is evolution to the modern mind? The *Oxford Dictionary of Biology* gives this definition: "The gradual process by which the present diversity of plant and animal life arose from the earliest and most primitive organisms, which is believed to have been continuing for at least the past 3000 million years." According to Denyse O'Leary, "evolution is the theory that all life forms are descended from one or several common ancestors that were present on the early Earth, three to four billion years ago."[5] And Michael Behe writes: "In its full-throated, biological sense, evolution means a process whereby life arose from non-living matter and subsequently developed entirely by natural means."[6]

Darwinism and 'Darwinism'

It is erroneous to associate evolution exclusively with Charles Darwin, although proclamations that "we live in the age of Charles Darwin" and comparisons of Darwin with Copernicus, Newton or Einstein are rife in the popularization of science as divulged by the media. To avoid being controversial they lean heavily on the tenets of scientific materialism, back up this official science with their own hyperbole, and if they occasionally serve up unorthodox items, it is with a sauce of denigrating irony.

All the same: "Darwin is not a strict Darwinian," and the man who said so was none other than a renowned self-proclaimed Darwinian, Stephen Jay Gould.[7] In fact, as we will see, Darwin was anything but a strict Darwinian in the current sense. "Darwin must be distinguished from modern Darwinism. One of the primary justifications for examining Darwin's own views is precisely to expose the frequent mismatches between the Darwin who is invoked by today's biologists eager to defend their corner, and the Darwin who wrote *The Origin of Species* and *The Descent of Man*."[8]

Nor was or is Darwinism the sole evolutionary theory. There is e.g.

---

4. Sri Aurobindo: *Essays Divine and Human,* p. 385.
5. Denyse O'Leary: *By Design or by Chance?* p. 9.
6. Michael Behe: *Darwin's Black Box,* p. xi.
7. John Brockman: *The Third Culture,* p. 53.
8. Tim Lewens: *Darwin,* p. 6.

Lamarckism, far from defunct although often so pronounced; vitalism, taboo in academe but stubbornly raising its head time and again in various disguises (evolution, after all, is about life); there is the 'Omega Point' theory of Pierre Teilhard de Chardin and his epigones; there is, of course, creationism, not only as narrated in some holy books, but also in its metaphorical variations; there is the intelligent design theory, which posits that the complexity in nature can only have been fashioned by a special Intelligence; and there is the very scientific but fiercely opposed theory of 'punctuated equilibrium' formulated by Stephen Gould and Niles Eldredge. This enumeration is far from complete but will have to do for the moment. The situation is actually such that, some say, every evolutionary biologist has his own theory.

Then what is authentic Darwinism? It is the theory gropingly worked out by Charles Darwin (1809-1882), especially in two of his books: *The Origin of Species by Means of Natural Selection, or the Preservation of Favoured Races in the Struggle for Life* (1859), and *The Descent of Man* (1872). The first book has had an enormous impact on the way humanity came to see itself. It has become the bible of present-day Darwinian biology and is quoted with reverence. To this end it is very useful, as in some cases Darwin has defended both sides of his arguments. "Before Darwin's death in 1882 the *Origin* went through a further five, often substantially revised, editions. They were revised so frequently and so radically because Darwin had found it increasingly difficult to deal with the problems presented to him by some of his more acute critics."[9]

Given this reverence even by such rational and critical-minded people as positivist scientists, it is amazing to what degree Charles Darwin has been exalted, not to say canonized. "Reading Darwinist literature, one cannot help noticing the way in which each writer stresses his or her own orthodoxy and total fidelity to Darwin, much like bishops discussing the encyclicals of a pope."[10] Novelist Barbara Kingsolver describes Darwin's idea of natural selection as "the greatest, simplest, most elegant logical construct ever to dawn across our curiosity about the workings of natural life. It is inarguable, and it explains everything."[11] And the philosopher Daniel Dennett credits Darwin with "the single best idea anyone has ever had."[12]

What, then, is 'Darwinism'? It is an agglomerate of theories

---

9. Derek Gjertsen: *Science and Philosophy,* p. 106.
10. Denyse O'Leary: op. cit., p. 111.
11. William Dembski (ed.): *Uncommon Dissent,* p. xvii.
12. id., p. xxii.

assembled and frequently revised under an umbrella postulated to be Charles Darwin's original idea. It is this cluster of more or less integrated theories which, after Darwin and up to the present, claims to prove that Darwin's evolutionary machinery would have been working, though he lacked most of the parts. Gregor Mendel's theory is among the best known; August Weismann's work is familiar only to experts, although it was defining for 'Darwinism'; and then came Hugo de Vries who introduced the mutations, Thomas Morgan and the application of mathematics to biology, 'the new synthesis' also called 'neo-Darwinism', the discovery of the double helix as the structure of DNA, sociobiology, and the theory of 'punctuated equilibrium', to name the most important elements of the cluster.

Darwin knew nothing of the 'mechanisms' which could explain his variation, natural selection or inheritance. The composition of the cell, the chromosome and the gene would be discovered decades later; even the fertilization of the ovum by the spermatozoon was still a mystery; and the discovery of the double helix would have to wait till 1953, a century after the publication of the *Origin*. "Darwin's scientific arguments are extremely weak, quite simply because in 1859 one was still completely ignorant of the mechanisms of reproduction and heredity," writes André Pichot, a historian of science.[13] Claude Allègre agrees: "Darwin's book contains indeed little proof. His book consists for the most part of conjectures, because he did not have the essential elements at his disposal to establish his theory."[14] Darwin himself had admitted something similar in a letter: "It deserves especial notice that the more important objections [to his theory] relate to questions of which we are confessedly ignorant; nor do we know how ignorant we are."[15]

A Sea-Going Naturalist

Charles Darwin was born in a rich family and could look forward to a life without financial worries. His grandfather and father were doctors, and Charles was expected to become likewise. He was therefore sent to the University of Edinburgh, the citadel of things medical in Britain. But young Charles could not stand the bloody slaughter on the

---

13. André Pichot: *Histoire de la notion de vie*, p. 808.
14. Claude Allègre: op. cit., p. 132.
15. Charles Darwin: *The Origin of Species,* p 431.

dissection table which, at a time that anaesthetics were unknown, was a dreadful affair. What Charles did feel attracted to was nature and her landscapes, plants and animals. He was instinctively drawn to people like the zoologist Robert Grant, the first to talk to him enthusiastically about evolution and its explanations, in this case the theory of the Frenchman Lamarck.

As Charles did not have the character to become a medical practitioner, his father allowed him to join Trinity College, at Cambridge University, in order to study theology and become a member of the Anglican clergy. A college education was a *sine qua non* for a clergyman; this made the universities of Cambridge and Oxford the main breeding ground of the official religion. The prospect of a sinecure as a country clergyman, a vicar, attracted Charles, for then he would have lots of leisure to devote to his hobby of exploring nature. At that time many of the books on "natural science" were indeed written by clergymen. "Naturalism was mostly the preserve of enthusiastic amateurs: clergymen whiling away idle moments in their rural parishes, and genteel young women drawing butterflies and pressing plants ..."[16] – "I believe that I was considered by all my masters and by my father as a very ordinary boy, rather below the common standard in intellect," wrote Darwin in his *Autobiography*. "To my deep mortification my father once said to me: 'You care for nothing but shooting dogs and rat-catching, and you will be a disgrace to yourself and all your family.'" Yet Destiny had other prospects for him.

When he was twenty-two, he received a proposal for a voyage on HMS *Beagle,* a 90-foot coaster of the Royal Navy preparing for a surveying mission. It had to test a new generation of clocks, of prime importance for calculating a ship's position, and to map the coastlines of South America. The captain, Robert FitzRoy, himself only twenty-seven but highly qualified, was looking for a gentleman who would be a suitable companion and who at the same time could make himself useful on his ship. Darwin, with his intense interest in nature, could be the ship's 'naturalist'; and as FitzRoy was a rather fanatical Bible reader, Darwin's theological studies would fit in nicely also.

To carry a naturalist, i.e. a person who studied nature, was normal procedure on a mission like this. Much of the planet and its denizens was still unknown, and the new knowledge favoured the expansion of the British Empire, the trade of its merchants and the zealous efforts

---

16. Iain McCalman: *Darwin's Armada,* p. 21.

of its missionaries. The nineteenth century was the age of geological, zoological and botanical exploration, of which the story of the mutiny on the *Bounty* is a telling illustration. Normally it was the ship's chief surgeon who doubled as naturalist, with the assistant-surgeon, in most cases a young physician, as his helper. It was as assistant-surgeon that the botanist Joseph Hooker travelled to Antarctica on the *Erebus* under Captain Ross, and that Thomas Huxley visited Australia and the surrounding region aboard the *Rattlesnake.*

This made Darwin's position on the *Beagle* quite exceptional, for he was not a surgeon but a sort of unqualified gentleman who paid for most of the expenses from his own purse. No wonder, then, that the chief-surgeon and actual naturalist, Robert McCormick, felt threatened by Darwin's status as confidant of the captain and by his untiring activities as a collector and scientist. McCormick "would leave the ship at Rio, cursing FitzRoy for allowing an unqualified outsider to usurp his domain. Darwin thought it a good riddance: the man was a pompous ass with antiquated ideas."[17]

HMS *Beagle* left Portsmouth in the last days of the year 1831 for a voyage around the world which would last five years. While the ship cruised down the east coast and up the west coast of South America, Darwin made several excursions inland. His voyage became one of discovery without end. He sent crates full of strange insects, gigantic fossil bones from unknown monsters, plants, birds and other animals to the motherland, where they were in eager demand for private collections and the first museums. He witnessed an eruption of a volcano, Mt. Osorno in Chili, and marveled at its titanic power to change the aspect of the Earth. And he was puzzled by the fact that, on an archipelago of small islands like the Galapagos, animals of the same species could show such marked differences. His "burning zeal to add even the most humble contribution to the noble structure of natural science" still increased, and could not be tempered by the constant sea-sickness which would affect his health for the rest of his life.

From the Galapagos the *Beagle* set sail for Australia. Having dropped anchor in Sidney, "my first feeling was to congratulate myself that I was born an Englishman."[18] And nearly two years later, after rounding Cape of Good Hope, Darwin set foot again on his native soil, in October 1836. This was no longer the naive nature-lover of five years earlier. He

---

17. Iain McCalman: op. cit., p. 45.
18. Tim Lewens: op. cit., p. 23.

had become rated as one of the experienced and knowledgeable world-explorers for the natural sciences, and the narrative of his eventful voyage on the *Beagle,* based on his diaries, rendered him acceptable in scientific circles. "The great amateur"[19] had arrived.

He would never go on another journey in his life.

Evolution in the Air

"Precisely when Darwin came to believe in evolution, whether it was a gradual dawning or a sudden realization, we will probably never know." (Michael Denton[20]) What we already know is that young Charles was driven by a "burning zeal" to add his personal contribution to natural science. He also thought of himself as a "philosophical naturalist", which meant "a naturalist whose classifications should not merely fit the pragmatic purpose of recording observations, but one who looks to give some rationale for nature's mode of organizations. More specifically, the rationale should be based on natural laws."[21] In other words, Darwin felt the urge not only to gather objects, facts and phenomena, but to look for explanations behind them and make them fit together in a theory.

There existed already a number of evolutionary theories when Darwin formulated his. The French *philosophes,* champions of the Enlightenment, had proposed several solutions to the transformations in the world of life, and Denis Diderot had summarized them in his influential but officially proscribed *Encyclopédie.* Buffon (1707-88) and Georges Cuvier (1769-1832) were learned and widely respected scientists with strong opinions for and against evolution. But the profoundest influence went out from the classification of all (then known) natural things by Carolus Linnaeus, the Latinized name of the Swedish botanist Karl von Linné (1707-78).

Linnaeus was, like everyone at the time, familiar with the Chain of Being, an age-old order of existence in a hierarchy of increasing complexity and consciousness: minerals, plants, animals, and at the top the lord of creation, the human being. But in his *Systema Naturae* (1735) he undertook the daring step to include the humans into his classification

19. Hilary and Steven Rose: *Alas Poor Darwin,* p. 111.
20. Michael Denton: *Evolution – A Theory in Crisis,* p. 27.
21. Tim Lewens: op. cit., p. 74.

of nature, still at the top, yes, but all the same in the company of the animals. To this end he created the class of the 'primates', containing the monkeys, the apes and ... the human beings. Because he never openly put the Christian creation myth in doubt, Linnaeus became a much honoured scientist in his own country and in the rest of Europe. He became known as "God's Registrar", and it was said that *Deus creavit, Linnaeus disposuit*: God created, Linnaeus classified.[22] The inclusion in his classification of the human species, *Homo sapiens,* would be essential in the evolutionary theories to come, and his influence, though rarely acknowledged, has ever been of the essence.

A direct precursor of Charles Darwin was his own grandfather, Erasmus Darwin (1731-1802). Erasmus was an excellent doctor, invited by King George III to be his personal physician – an offer which the doctor declined. He was also a man of the broadest interests, a freethinker and "unabashed materialist", and as such "the very embodiment of enlightened values."[23] He constructed his own comprehensive theory of evolution. "He reasoned that life had not been created in the Garden of Eden but had arisen naturally and gradually, by stages, from the most elemental microscopic stuff."[24] His thought ran along the same lines as Lamarck's: species changed by adaptation to their environment. Charles Darwin certainly read his grandfather's *Zoonomia or the Laws of Organic Life* (1796). Yet, in his *Origin of Species* he dismissed him in a footnote as "a pre-Lamarckian harbinger of Lamarck's confusion. And in his *Autobiography* Darwin spoke disparagingly of Erasmus's *Zoonomia,* the book that may well have planted in Darwin's mind the seed not only of evolutionism, but of the theory of natural selection."[25]

One may be amazed to learn that Darwin was a Lamarckian himself, a fact which has long been disclaimed by Darwinian authors but which they can no longer deny. The Darwin-versus-Lamarck controversy has been one of the main features of evolutionary biology. "The caricature of Lamarck's position that we have inherited today" can still be found, whenever he is mentioned, in curt negations of his historical importance, like: "Lamarck believed in all sorts of things that have been rejected." Statements of this kind usually betray an un-scientific attitude or a lack of knowledge. It is a frequent experience for the student of the history and philosophy of science to find how reputed authors

---

22. Michael Shermer: *In Darwin's Shadow,* p. 101.
23. Roy Porter: *Flesh in the Age of Reason,* p. 377.
24. Roy Porter: op. cit., p. 382.
25. Robert Wright: *The Moral Animal,* p. 276

blindly copy incorrect matters or references, and thereby contribute to the creation of untruths and outright legends. André Pichot does have reasons to write: "The history of the biology of the last two centuries has been altered by numerous legends."[26]

Lamarck (1744-1829) was the first great evolutionary theorist. Goulven Laurent, in "The Birth of Transformism – Lamarck between Linné and Darwin,"[27] calls him "the French Linnaeus". He attributes to him the classification of the invertebrate animals, the founding of their paleontology, the promotion of the concept of biology (a word he coined), the introduction of the word 'fossil' in its present sense, and the formulation of 'transformism', the term then used by French scientists for what we call 'evolution'. Lamarck based his view on pure materialism (although, as a good deist, he recognized "a sublime Author of all things").

His theory of 'transformism' rested on two principles: organs are created by the need and use of them, and acquired characteristics can be inherited. This second principle means that change in the species could take place from the outside inwards: the changes in a living being during its life could be transmitted to its offspring. Darwin, on the contrary, held that inherited changes were the result of small variations within the bodies of a species which resulted themselves gradually in changes on the outside. Neither of them had an explanation for the mechanism of their 'transformations'. (Darwin himself did not use the word 'evolution' in his *Origin*, he called it 'transmutationism'.) Science had not yet progressed sufficiently to allow an understanding of what were, in sum, guesses based on research and experience. This lacuna made Darwin, conscious of the fragility of his position, defend his theory all the more tenaciously, not to say desperately, and may explain his denigrating remark about Lamarck quoted above.

The astonishing truth is that Darwin was as much a Lamarckian as his French predecessor! "It is Darwin who has used for the first time the term 'inheritance of acquired characteristics', unknown to Lamarck, and who has tried to make it into a theoretical justification for his 'pangenesis'."[28] Lamarck had never presented a theory to support the inheritance of acquired characteristics; it was at the time the consensus among scientists, and it would remain so throughout Darwin's life. Darwin, however, concocted a theory of pangenesis and 'gemmules',

---

26. André Pichot: op. cit., p. 860.
27. Goulven Laurent: *La naissance du transformisme – Lamarck entre Linné et Darwin.*
28. Goulven Laurent: op. cit., p. 131.

secreted internally by each part of the body, gathered in the sex cells, and transmitted to the embryo through the union of sperm and egg. This idea was later discarded as one of his blunders and fell into oblivion. But "the Darwinian heredity of acquired characteristics is clearly affirmed in several places of *The Origin of Species*, and even, rather curiously, in a fully Lamarckian manner."[29] It was the reason that Darwin grew obsessed with the fear that his children might have inherited the illness he had contracted on the *Beagle,* which was impossible according to his own theory of variation and natural selection, so often held to be his only and definitive view.

---

29. André Pichot: op. cit., p. 812.

2.

The Making of a Theory

It is the customary fate of new truths to begin as heresies and to end as superstitions.

Thomas H. Huxley

The Smile of the Cheshire Cat

No sound escapes the ear of a soul touched by the Muse of music, no shade of colour the eye awakened by the Muse of art – and not a blade of grass or scrap of information about nature the budding 'naturalist'. Charles Darwin drank it all. Evolution was in the air, and we know how he had listened when Robert Grant expounded animatedly Lamarck's theory of evolution. Darwin talked with any expert he met and read anything about nature he could lay his hands on. He recalls in his *Autobiography* how inspired he was by the narrative of Alexander von Humboldt, the German naturalist and explorer, about his discoveries in Central and South America.

There was a sea-change taking place in Britain's 19th century. The Enlightenment is generally associated with France and the *philosophes*, but the part of the British philosophers, Thomas Hobbes, John Locke and David Hume in particular, was at least as important. Their critical attitude (empiricism) towards the long established European, mainly Christian values opened the gates for the principles of materialism, atheism, liberalism, science and progress. Most of the British intelligentsia had been shocked by the French Revolution, but even in Britain the rise of the new ideas could not be halted. The result was the Victorian age, named after the dapper and long-reigning Queen Victoria (1819-1901), and symbolizing a period of official Anglican religion (similar to Catholicism), strict morals and customs, and behind that dignified façade all intellectual viewpoints in as well as out of fashion.

Destined to be a clergyman, Charles Darwin was nevertheless

inevitably influenced by materialism and atheism. His grandfather, father, brother, and almost all his best friends were disbelievers, freethinkers who did not hide their convictions. Moreover, the sciences had become materialistic and mechanistic since Galileo Galilei, René Descartes, Isaac Newton, and Laplace. God in his heaven had still been reverenced for some time, though mainly to avoid the persecution and heavy punishment Giordano Bruno, Galileo and many others had suffered. In Britain, for example, "the bishops considered Hobbes to be an enemy of the church and the most dangerous atheist in England. ... In 1666, bishops blamed the fire [in London] on his atheism. Parliament investigated him for blasphemy for two years ... He escaped being charged as a heretic, but he was forbidden to write ever again about human nature. ... Locke worried that his writings might get him hanged."[1]

Darwin knew of more recent cases and therefore "hid his materialism, as he secretly scribbled away in his notebooks working out his theory."[2] As he wanted to be a "philosophical naturalist" and contribute to the science of his time, he was forced to take an increasingly critical stance towards the religion of his youth, and of his wife. On the *Beagle* he had been "quite orthodox, but his faith waned with time ... The Old Testament was 'no more to be trusted than the sacred books of the Hindoos [*sic*], or the beliefs of any barbarian'."[3] In time he could no longer accept miracles, hell, and the suffering in nature ordained by a good God. The death of his favourite daughter, Anne, at the age of nine, delivered the final blow to his formerly cherished religious convictions.

The more he thought, the more bewildered he became. Later in life he will describe himself as an "agnostic," somebody who does neither accept nor deny the existence of God, and who acquiesces in his ignorance. In the mind of his century and the previous one, the belief in God's existence had faded from self-evident to a rationally accepted opinion, then to doubt, and finally to denial. Nietzsche's Zarathustra will proclaim the death of the Christian God. As the historian A.N. Wilson writes, God had disappeared like, in *Alice in Wonderland,* the smile of the Cheshire Cat.

---

1. Carl Zimmer: *Soul Made Flesh – The Discovery of the Brain,* pp. 232-33, 249.
2. John Forster e.a.: *Critique of Intelligent Design,* p. 111.
3. Tim Lewens: *Darwin,* p. 35.

Darwin's Conversion

Darwin's scientific thinking followed a development parallel with that of his changing religious beliefs. Since Descartes the dominant scientific view of the world had become mechanistic. Descartes had still accepted a rational soul, but held that the body and all animal life forms were machines. Newton had been an alchemist and a Bible exegete, but his model of the universe was a kind of clockwork, put in motion by God Almighty, but afterwards left to its own automatic movement. And Laplace, in a legendary answer to Napoleon, had declared that he no longer needed the hypothesis of a God to explain the workings of the universe.

As a reaction against this increasing materialism, religiously inclined persons, in the first place learned members of the clergy, launched 'natural theology'. One of the early proponents of this view was John Ray, the title of whose book *The Wisdom of God Manifested in the Works of Creation* (1691) states in a nutshell what natural theology was about. Ray was the first to use the metaphor of the watch: if you happened to find in nature something made like a watch, you could not but deduce that it must be made by an intelligent being, a watchmaker. Therefore, seeing how marvellously everything in nature and the universe had been put together and was functioning, one could not but deduce that there was a supreme Intelligent Being who had made it all. You had to conclude that God existed.

At Cambridge William Paley's book: *Natural Theology – or Evidence of the Existence and Attributes of the Deity Collected from the Appearances of Nature* (1803) had been mandatory reading. Darwin had been fascinated by this and other works of Paley, one of the last writers on the subject. "The careful study of these works was the only part of the Academical Course which was of the least use to me in the education of my mind. I did not at that time trouble myself with Paley's premises; and taking these on trust I was charmed and convinced by the long line of argumentation."[4] Darwin had boarded the *Beagle* as a convinced natural theologian.

A fixed mental make-up of this kind can only be transformed by a long and stressful process of conversion. The most persuasive indications that species can and do change must have come from the plant and animal breeders whom Darwin interrogated on every possible occasion,

---

4. Charles Darwin: *Autobiography,* pp. 30-31.

and possibly from the fossil record as it was then known, sustained by the Linnean classification. While his idea of the evolutionary process gradually took shape, he continued the studies which contributed to his increasing reputation as a naturalist. He studied coral reefs, climbing plants, earthworms and orchids, and dissected barnacles for a full eight years, from 1846 till 1854. Then: "At least gleams of light have come, and I am almost convinced (quite contrary to the opinion I started with) that species are not (it is like confessing a murder) immutable."[5]

He deemed Paley's old argument of design in nature, "which formerly seemed to me so conclusive," no longer valid, for he had thought out a scientific one. He had become convinced that all species originated from a common source, like the branches of one tree. The changes, or "modifications," into different branches were caused by variations in the individuals of the species, the existence of which was confirmed by every breeder. The ruthless competition for food in the natural environment would allow only the strongest individuals to survive, in other words the carriers of the variations most fit to adapt. And over time these fittest animals would transmit their physical adaptations to their offspring.

It would still take Darwin some time and one or two lucky breaks to formulate his theory in this way, but it had become clear to him that Paley's religious view belonged to the past. Darwin could, in his time and position, only come forward with a materialistic and atheistic explication. Species changed; the fossils showed that many had died out; the chain of being (never referred to but present in the background) and the new geology showed that changes took place gradually over long periods of time; the fight for survival was a common fact of existence. How these mechanisms of life actually worked, Darwin had as yet no idea, but future science would certainly find out. Like Laplace he neither needed a God anymore.

Gradually, Gradually

Darwin had met Charles Lyell (1797-1875) personally and had felt much honoured by their acquaintance, for the slightly older geologist was already a person of esteem and Darwin still a nobody. Professor Lyell was building on the theories of James Hutton, the founder

---

5. Michael Denton: *Evolution – A Theory in Crisis,* p. 34.

of modern geology, who in 1795 "believed that the surface features of the Earth were shaped gradually by incremental changes extending over enormous lengths of time. He realized that millions of years would be needed to accumulate rock sediments and to raise and erode mountains."[6]

In Darwin's days the first trains were on the rails, but the idea of time was still based on interpretations of the Bible. In 1620 Archbishop James Ussher, "who had laboured over his studies for decades, even when he became chaplain to the king of England," had concluded that Adam was created at 9 a.m. on Sunday, 23 October 4004 BC. Time had began on the previous day at 6 p.m. Nowadays one may react to declarations like these with a smile, but "Ussher's date was recognized by the Church of England in 1701, and was thereafter published in the opening margin of the King James Bible right the way through to the twentieth century. Even scientists and philosophers were happy to accept Ussher's date well into the nineteenth century."[7] "In the early nineteenth century even Charles Darwin would graduate from Cambridge University believing that the world was six thousand years old, give or take."[8]

As Bertrand Russell reminded us: "Nobody nowadays believes that the world was created in 4004 BC; but not so very long ago scepticism on this point was thought an abominable crime."[9] It is worth reflecting on this fact to comprehend how much the world of Charles Darwin, in which his theory of evolution was conceived, differed from ours. The editor of *The Faber Book of Science* writes: "The remote antiquity of man was popularized not so much by a theory [i.e. Darwin's] as by the discovery of a vast and undeniable subject matter: a new dark continent of time, prehistory. More persuasively than a theory, the artifacts themselves seemed to bear witness to a chronology of prehistory ... Gradually the idea entered popular consciousness ... In 1867, the announcement of the first *Congrès international préhistorique de Paris* brought the first official use of the word 'prehistoric'."[10]

The "disturbing notion" that man had existed long before 4004 BC, prudently proposed by scientists from Buffon onwards, was accepted with hesitation even by the scientific community. Such was the pressure

---

6. Paul Davies: *The Origin of Life*, p. 134.
7. Simon Singh: *Big Bang*, p. 77.
8. Stephen Baxter: *Revolutions in the Earth*, p. 17.
9. *Bertrand Russell on Religion*, p. 177.
10. John Carey (ed.): *The Faber Book of Science*, p. 136.

from the powerfully authoritative Church, supported by the State, that every breach or widening of the narrow, rigid circle of general aware-ness threatened its author with the danger of punishment and social discrimination. Innovation and progress have needed their Socrates, Luther, Bruno or Galileo in all cultures, in all times.

Gradualism in geology was initiated by Hutton, taken over by Lyell, and accepted by Darwin. The first geologists, in Britain as well as in France, went exploring what was nearest to them: their own neigh-bourhood, region, country. (This is the reason that so many British and French place names are found in their geological nomenclatures.) Although they were very perceptive and accurate in the descriptions of their field work – they were without exception passionate researchers, even when amateurs – what they were exploring was a geological situa-tion which had remained stable as long as they could remember. "Not even Lyell himself had seen a volcano erupt." And so it happened that Hutton, and Lyell after him, formulated the principle of 'gradualism', according to which all changes in the surface of the Earth were the result of small differences which had gradually come about over long periods of time.

"Lyell applied Hutton's great idea [of gradualism] with a ruthless severity that even Hutton would not have recognized. In the past, Lyell said, Earth had always looked much as it looks now ... Everything concerning plant and animal life was determined: similar conditions recurring in the future would give rise to *exactly* the same species as in the past ... Lyell's uniformitarianism [another name for gradualism] was taken to a fanatical degree. He insisted that every past event had to be explained by causes now operating: there was nothing in the past that we cannot see in the world around us now."[11] In science, this way of seeing is called 'extrapolation backwards'. Even common sense will tell that it is a slippery way of reasoning, because experience teaches that things are never exactly the same.

In 1830 Charles Lyell published the first volume of his trilogy: *Prin-ciples of Geology – Being an Attempt to Explain the Former Changes of the Earth's Surface by Reference to Causes now in Operation.* The book came just in time for Darwin to take it with him on the *Beagle,* where he had ample leisure to read and reread it. If ever he had hesitated between sud-den and gradual changes, Lyell convinced him of the latter. Darwin had witnessed a volcanic eruption and seen with his own eyes what massive

---

11. Stephen Baxter: op. cit., pp. 204, 206.

sudden upheavals it could cause; still he decided for gradualism. His decision, taken against the advice of Thomas Huxley and others, would have long-lasting consequences in the study of evolution, even at the present day.

A Struggle for Survival

Charles Darwin has been called "the midwife of the idea of evolution" and there is substance in this metaphor. He caught as many seeds as possible floating on the winds of the century, covered them with the humus of his imagination, and was finally rewarded with the plant of his theory. We have met with some of the influences on his thought, but there were many others. There was e.g. John Herschel's *Introduction to the Study of Natural Philosophy* (Darwin had met the astronomer, discoverer of Uranus, in Cape Town). And there was more recently Robert Chambers' *Vestiges of the Natural History of Creation*, a theory of evolution in the Lamarckian way, published anonymously for the sake of safety, and which had sparked a furore.

But the revelation that put everything together was caused by Thomas Malthus' *Essay on the Principle of Population as it affects the Future Improvement of Society,* first published in 1798. "Darwin would regard the reading of this book [in 1838] as an extended 'Eureka!' moment in the formulation of his views."[12] Thomas Malthus, another clergyman, was an economist and demographer. He seems to have been struck by the sight of the impoverished masses created by the industrial revolution of which Britain was the cradle. The presence of those proletarians was a new phenomenon in British society and the established classes felt extremely uncomfortable with them. (Marx and Engels wrote the *Communist Manifesto* in 1848.)

From his social perceptions Malthus concocted a theory: population increases geometrically, the available food increases only arithmetically. This meant that there would always be less food than people vying for it, and that famine, war and ill health were the only certainties in the future. Depraved existence in poverty was humanity's inescapable lot. "In Malthus' grim essay, Darwin found the engine that could push evolution forward."[13] As the life of humanity was destined to be a permanent

---

12. Tim Lewens: op. cit., p. 26.
13. Carl Zimmer: *Evolution,* p. 41.

struggle for survival, so must be the life of all beings in nature. Darwin applied the doctrine of Malthus to the animal and vegetable kingdoms.

He found his conclusions confirmed in "the writings of Adam Smith and other utilitarian economists who presented individual competition as the driving force of economic progress. Perhaps more important, he lived in a [capitalist] society that embraced this view."[14] This was the explanation for the improvement and survival of the fittest, the success of their superior personal characteristics, and ultimately the appearance of new species. "It at once struck me," wrote Darwin afterwards in a letter, "that under these circumstances favourable variations would tend to be preserved, and unfavourable ones to be destroyed. The result of this would be the formation of new species. Here, then, I had at last got a theory by which to work."[15]

Looking back on this chapter, we see that the pillars of Darwin's biological theory of evolution were in fact a one-sided geological theory, and a generalization of a socio-economical theory concerning a first-hand reaction to a new, unfamiliar and scary social situation. On biological facts an conceptual grid was projected that did touch some points of the reality, but left most others untouched and scientifically unexplained, for instance the mechanisms of inheritance and adaptation (which were still unexplainable at the time). The illumination which brought it all together in Darwin's mind, Malthus' theory, proved to be a chimera. As Tim Lewens writes in his book on Darwin: "Modern evolution has no essential commitment to the Malthusian view that lies at the heart of Darwin's theory."[16]

---

14. Edward Larson: *Evolution,* p. 70.
15. Michael Denton: op. cit., p. 42.
16. Tim Lewens: op. cit., p. 60.

3.

The Origin of Species

When the views advanced in this volume, and by Mr. Wallace, or when analogous views on the origin of species are generally admitted, we can dimly foresee that there will be a considerable revolution in natural history.

CHARLES DARWIN

Forced to Publish

Darwin felt that the interpretation of evolution he had worked out was "like confessing a murder." He knew that it would be understood at once that his theory put the human being on an equal footing with monkeys and apes, even if he postponed writing explicitly about this aspect of the question. That Frenchmen (Lamarck) spread this kind of nonsense was scandalous, though less so than cutting their King and Queen's heads off; but that an Englishman could utter such blasphemy and pretend to justify it was, to say the least, shocking. Darwin knew that publishing his theory would brand him publicly as a materialist and a freethinker, and that his beloved wife Emma would be hurt in her religious feelings. Therefore he delayed publishing the theory year after year, wrote furtively a summary of it somewhere along the way, and dissected countless barnacles, crustaceans resembling mussels which attach themselves to the ships' hulls in clusters. Which Christian would take it gracefully that there was a monkey among the ancestors of Jesus Christ?

But in June 1858 Darwin received an envelope from overseas containing a hastily scribbled essay which, to his stupefaction, developed exactly the same thesis as his. If the author "had read my manuscript sketch written in 1842 he could not have made a better short abstract!" Darwin wrote to Lyell in desperation. The author of the "twenty or so pages of text on rice paper" was Alfred Wallace (1823-1913), a former schoolteacher who had become a passionate naturalist, and was at that time exploring the Malay Archipelago. "I can have no fear of having to suffer for the study of nature and the search for truth," he had once written.

Darwin had been five years on the *Beagle*; Wallace had been four years in Amazonia and would remain a full seven years in Malaysia. He escaped grave danger and even death on many occasions, and contracted malaria. It was during a bout of fever that he, too, read Malthus' *Essay on Population,* and as he, too, had read Lyell's *Geology,* he, too, had a sudden illumination. "There suddenly flashed upon me the idea of the survival of the fittest ... The whole idea of specific modification [i.e. changes of the species] became clear to me, and in the two hours of my fit I had thought the main points of my theory."[1]

Charles Darwin had a problem. If he did not react and make his theory public without further delay, Wallace's essay might be sent to others, and Darwin would loose the priority of a thesis on which he had worked for so many years and which had become the centre of his existence. Darwin had become an established naturalist with influential friends; Wallace was still practically unknown and in a far-away place on the globe; he was "someone who was no gentleman of science, but an obscure butterfly-collector." The solution Darwin's friends came up with was "to read both Wallace's paper and Darwin's sketch of 1844, along with a letter Darwin had written to [the Harvard geologist] Asa Gray on 5 September 1857, outlining his ideas (and thus establishing priority under the rules of the time), at the 1 July 1858 meeting of the Linnean Society." "Wallace's co-discoverer status with Darwin's is generally accepted by all biologists and historians," writes Michael Shermer.[2]

Yet there still are some who doubt the fairness of the 1848 procedure, and others who suspect Darwin of plagiarism and tampering with the date of a letter. "The balance of probabilities exonerates Darwin ... And yet the miasma of conspiracy that hangs over the events of June-July [1848] is not entirely dispelled," finds Iain MacCalman.[3] Lyell and Hooker were protecting their personal friend and social equal [i.e. Darwin], and defending the ranks and procedures of respectable science against the lower-class outsider Alfred Wallace.

"Had Darwin's friends acted immorally? Certainly they had bent

---

1. Michael Shermer: *In Darwin's Shadow – The Life and Science of Alfred Russel Wallace,* p. 113.
2. id., pp. 119, 129.
3. id., pp. 322, 323.

the rules to advance their friend's position at Wallace's expense." In 1855 Wallace had published the "Sarawak Law paper", which was "the first ever British scientific paper to claim that animals had descended from a common ancestor and then produced closely similar variations which evolved into distinct species."[4] This paper gave Wallace priority of publication of his theory of evolution, but it was not mentioned in the documentation of the case as presented at the decisive meeting of the Linnean Society. Moreover, everything was done post-haste without consulting Wallace. That Wallace afterwards humbly agreed to the whole procedure is of course no proof that it had been fair.

By this affair Darwin was at last sufficiently motivated to begin at once to write what would become his most famous, world-changing book.

The Book

On the Origin of Species by Means of Natural Selection, or the Preservation of Favoured Races in the Struggle for Life was published on 24 November 1859.[5] A.N. Wilson writes that it sold in quantities to rival the novels of Charles Dickens. On the day the book became available, all 1,250 copies were already subscribed to by the retail trade, so it went immediately into a second printing. "Almost overnight one book transformed the scientific and popular debate over biological origins ... The staid voice of establishment in England, *The Times of London*, featured a wonderfully favourable review."[6] "Along with the Bible, *De Revolutionibus* [Copernicus], *Principia Mathematica* [Newton] and *Das Kapital* [Marx], Darwin's *Origin of Species* must rank among the least read (at least in full) and most influential books of all time. It is no exaggeration to talk of it bringing about a Darwinian revolution."[7]

"Today ensconced in our comfortable agnosticism, after a century [and a half] of exposure to the idea of evolution and quite inured to the idea of a universe without a purpose, we tend to forget just what a shock wave the advent of [Darwinian] evolution sent through the Christian society of Victorian England," writes Michael Denton. "Darwin's theory

4. Michael Shermer: op. cit., pp. 329, 263.
5. The edition used for reference in this book is the 6th, as published by Goyal Publishers & Distributors, Delhi (2006).
6. Edward Larson: *Evolution*, pp. 84, 79.
7. Francis Hitching: *The Neck of the Giraffe*, p. 249.

broke man's link with God ... Undoubtedly the most significant factor that contributed to the success of the Darwinian theory after 1859 was the fact that it was the first genuine attempt to bring the study of life on Earth fully into the conceptual sphere of science."[8]

Even after the impact of the first shockwave, though, "Christian commitment was not the exception but the rule ... Most of the leading scientists of Great Britain retained a Christian commitment." (A.N. Wilson[9]) In a thunderstorm lightning strikes only after clouds have gathered, and one strike does not exhaust the storm. Revolutions are a sudden discharge of built up tensions and bring about a drastic change, usually through much destruction, but they do not make the world completely different at once. A religious doctrine, a strong bulwark of the mind constructed and reinforced over centuries, does not crumble from the first bolt of lightning or the impact of a new idea, in this case Darwinism. The remnants of the bulwark of Christianity, built up during the Middle Ages, are still standing in the Western world. Darwinism was *one* of the phenomena of scientific rationalism trying to de-construct that bulwark – after the Renaissance, the Reformation, the scientific revolution in the seventeenth century, the Enlightenment, the Great Wars, the globalization, and the present confusion, presumably the end of the long process and a transition into a new world.

Darwin, midwife of the theory of evolution as understood in the common mind, has contributed to a change in the understanding of the world and humanity. "It was Darwin the symbol, Darwin the name which stood for a process, the name which was hurled from one side to the other in the polemics of secularist platforms or journals, an imaginary Darwin, a vague Darwin, without the comfortable homely substantial outlines of the real naturalist of a Kentish village, but how-ever imaginary and however vague still bearing a direct relationship to a scientific achievement which few quite understood, the truth of which many doubted, but which everyone, without quite knowing what it was, knew to be a scientific achievement of the first magnitude." (Owen Chadwick[10])

---

8. Michael Denton: *Evolution – A Theory in Crisis,* p. 66, 67, 71.
9. A.N. Wilson: *God's Funeral,* p. 241-42.
10. Owen Chadwick: *The Secularization of the European Mind in the Nineteenth Century,* p. 174.

The Darwin Sect

A new idea has to assert itself aggressively to stay alive and develop. Four in the vanguard of the defence of Darwin's idea were Thomas Huxley, Charles Lyell, Joseph Hooker and Asa Gray. They were known as "the four musketeers." Huxley, the fearsome fighter and passionate supporter of scientific naturalism, was called "Darwin's bulldog" and "the Apostle of Unbelief," which needs no clarification. Asa Gray was an American botanist and professor of natural history at Harvard University. He was one of the few persons Darwin kept informed about the progress of his thought and the publication of *The Origin of Species.* Like Lyell, Gray "staunchly supported" natural selection as the cause of new species, but contended, as a Christian, that the process was directed by providential influence rather than pure chance. Joseph Hooker, assistant director (later director) of the Royal Botanical Gardens at Kew, was a close friend of Darwin. One may remember that it was he, together with Lyell, who organized the historic meeting of the Linnean Society where the priority of the evolution theory was adjudicated to Darwin.

It is intriguing that all four doubted, and even opposed, any theory of evolution, including the one of their friend Charles Darwin shortly before the publication of the *Origin.* Whether to Christian, atheist or agnostic, a theory of evolution, advancing that human beings descended from monkeys and ultimately from invisibly small life-forms, microbes, seemed at the time utterly unbelievable – even to naturalists who spend their days studying living beings, and trying to find explanations for their fantastic diversity, beauty and sometimes horror or apparent cruelty.

"As a passionate supporter of scientific naturalism, Huxley felt bound to lash evolutionary thinking as bad science. … All theories of development were, in Huxley's mind, the product of metaphysical rather than naturalistic thinking."[11] Lyell, in spite of his support for Darwin in 1858, will convert to his theory only ten years later. Gray will only accept a Christian variation of the theory. And even Hooker, Darwin's closest friend, will remain doubtful till the book was published, despite having already mentioned the theory in previous writings.

André Pichot writes that Darwinism became aggressively anti-religious. McCalman's recent book *Darwin's Armada* confirms this contention in a surprising way. Darwin, Hooker and Huxley had, all

---

11. Iain McCalman: *Darwin's Armada,* p. 307.

three of them, been sea-going explorers for years and thus shared, in Huxley's words, "a masonic bond in being well salted in early life." Huxley "longed for science to be seen as a moral calling greater than any religion. ... He and Hooker had effectively become joint leaders of a group of young scientists who wanted to reform Britain's old guard of clerical dilettantism and entitlement." They wanted to "swamp the parsons ... to split science from theology." McCalman writes about Huxley's "anticlerical ferocity" and quotes him as saying: "If I have a wish to live thirty years [more], it is that I see the foot of Science on the necks of her enemies."[12]

Once they were converted and obliged to declare their positions when the *Origin* was published, Darwin encouraged his polemical friends in subtle ways. He began e.g. to speak of "our side" and to congratulate them on being "a good and compact body." The exceptional cohesion of the small group of his followers began to resemble a scientific sect and to function as one. This rarely mentioned aspect of the history of Darwinism became still more outspoken by the foundation, at the initiative of Huxley and Hooker, of the "X Club", which was thought of as a caucus or ring. Their aim was "to substantiate Darwinian evolution and turn Victorian Britain into a scientific [and atheistic] society."

"Huxley wanted the introduction of a scholarly and ethical form of science, open to merit; he wanted the overthrow of the clergy, aristocrats, and social climbers; and he wanted to use Darwinism as a Whitworth gun, to bring about a 'New Reformation' that would sweep away 'the scum of rotten hypocritical conventionalists which clogs art, literature, science and politics'." This was no less than a cultural coup which would profoundly mark the future of the biological sciences, and continues to do so today. They were "a meritocratic conspiracy" which drew its grit from the struggles Huxley and Hooker had had to fight in their early lives. "Collectively they were unstoppable. They nominated each other for awards, refereed each other for jobs, published each other's work, sponsored each other's lecture tours, awarded each other grants, and circulated each other's achievements."[13]

The feat Huxley is best remembered for, his confrontation in 1860 with Anglican bishop Samuel Wilberforce during a meeting at Oxford, should be seen against this background. According to the traditional account of that confrontation Wilberforce attacked the new evolutionary

---

12. Iain McCalman: op. cit., part five "The Armada at War," p. 293 ff.
13. id., pp. 347, 356.

theory in a speech, and ended with the question whether Darwin was related to the apes on his grandfather's or grandmother's side. Huxley claimed to have replied: "If this question were put to me, I would rather have a miserable ape for a grandfather than a man highly endowed by nature and possessed of great means and influence, but who employs those faculties for the mere purpose of introducing ridicule into a grave scientific discussion – I unhesitatingly affirm my preference for the ape."[14]

The argumentative power of this reply is weak and its historicity doubtful. Wilberforce was a learned man whose 17,000 words critique of the *Origin* was published in a prestigious quarterly a few days after the debate. "What really happened [at that meeting] receded behind a fog bank of embellishment. Each of the players in the drama offered his own version in which he came best," writes Carl Zimmer.[15] For Joseph Hooker was also present and would write to Darwin "that he himself had taken on Wilberforce: 'I smacked him amid rounds of applause.' Both Huxley and Hooker were telling stories that would do two things: raise their stature in Darwin's eyes, and leave him indebted to them."[16]

And what about the title of the famous book on the origin of species? "Darwin was not able to present a single instance of speciation by natural selection in *The Origin of Species*." There were plenty of examples of artificial selection by dog- and pigeon-breeders, in which the human breeders provided the selecting intelligence. But even then "Darwin had to admit, when challenged, that he could provide no cases of animal-breeders producing a new species. Such breeding had definitely produced different varieties, but not a single new species."[17] As the authoritative Ernst Mayr states concisely: "Darwin himself failed to solve the problem of speciation."[18] – "If there is one thing *The Origin of Species* is not about," affirms to Steve Jones, "it is the origin of species."[19]

---

14. See e.g. Edward Larson: op. cit., pp. 93-94.
15. Carl Zimmer: *Evolution*, p. 66.
16. Robert Wright: *The Moral Animal*, p. 281.
17. Daniel Dennett: *Darwin's Dangerous Idea*, p. 101.
18. Ernst Mayr: *What Evolution is*, p. 193.
19. Daniel Dennett: *Darwin's Dangerous Idea.*, p. 44.

What Darwin Really Said

At the time Charles Darwin composed his theory genes were not known, the cell was a blob of jelly called 'plasm', dinosaurs did not roam the landscape of popular culture, and (the white-skinned) *Homo sapiens* still ranked at the top of the tree of life. Darwin's Darwinism was very different from the 'Darwinism' of the neo-Darwinists some eighty or a hundred years later, now the basis of evolution as presented by the media, and therefore commonly supposed to be the thought of the great man himself.

We have seen that Darwin's conversion was a turning away from his religious belief towards an agnostic materialism. Instead of the mystic vision of natural theology, he wanted an explanation of the species based on science. "Undoubtedly the most significant factor that contributed to the success of Darwinian theory after 1859 was the fact that it was the first genuine attempt to bring the study of life on Earth fully into the conceptual sphere of science," notes Michael Denton.[20] And Darwin himself wrote: "I had two distinct objects in view: firstly to show that species had not been separately created [as taught by the Christian religion], and second, that natural selection [a scientific mechanism] had been the chief agent of change."[21]

When the Renaissance had made the return of science possible – *la nuova scienza* – Galileo Galilei picked up where the ancient Greek scientists had had to leave off. He defined the principles of science which are still valid today. 1. Science should be about matter. This is now so self-evident that it is seldom a point of consideration. 2. Science has no grasp of wholes, but has to reduce all things to parts consisting of smaller parts consisting of still smaller parts. 3. All changes in matter are brought about by external forces. This excludes any kind of internal movement or life. 4. Science can only work with the 'primary' qualities of things: extension, motion, and mass. 'Secondary' qualities, like colour, scent or taste, are effects of the primary qualities. 5. The language of science is mathematics, based on measurement. 6. In science all guesses, theses, or theories have to be tested as to their truth and reality. The time of fantasies and superstitions belongs to the past. The foundation of the New Science is "the scientific method" and its key-procedure

---

20. Michael Denton: *Evolution – A Theory in Crisis*, p. 71.
21. Michael Behe e.a.: *Science and Evidence for Design*, p. 180.

the experiment. The elements of this enumeration have been, point for point, of immense importance in the latest four centuries.

It was within this framework of science that Darwin had to shift his reasoning from a world as perceived by natural theology. Anything he had been and was studying since his voyage on the *Beagle* was intuitively weighed against these premises. The wondrous results of breeding, which made a chihuahua from a wolf, were one of his strongest arguments. "The variation of domestic animals provided Darwin not only with evidence of the power of selection but also with irrefutable evidence that organisms could indeed undergo a considerable degree of evolutionary change." Another argument was the at the time still scarce paleonthological knowledge: "Paleontology also provided Darwin with evidence that evolution had occurred. The fossil record revealed that the history of life on Earth was overall one of progress from simple to more complex types of life."[22]

Then what did Darwin actually say? According to André Pichot the Darwinian thesis can be summarized as follows: "1. More living beings are born than the natural resources are able to feed. ['This', comments Pichot, 'is the thesis of Malthus, but considerably expanded and applied to the whole of the animal kingdom.'] 2. The various members of a species show very slight individual differences – for reasons which are not explained. 3. The beings which improve their individual differences have the advantage in the competition for the natural resources. 4. Accumulation of the slight differences, selected in this way, results in the creation of new species."[23] – "In essence," writes Michael Denton, "the scientific mechanism [proposed by Darwin] depended on only three premises each of which were practically self-evident: that organisms varied; that these variations could be inherited; and that all organisms were subject to an intense struggle for existence which was bound to favour the preservation by natural selection of beneficial variations."[24]

"Never were so many facts explained by so few assumptions," marvels Richard Dawkins,[25] who labels himself an "arch-Darwinist." Pichot is less impressed and finds "Darwin's thesis rather clumsy and hardly coherent."[26] Claude Allègre concurs: "Darwin's book proves indeed very little. For the most part it rests on conjectures, as Darwin does

---

22. Michael Denton: op. cit., pp. 45, 52.
23. André Pichot: *Histoire de la notion de vie*, p. 790.
24. Michael Denton: op. cit., p. 42.
25. Richard Dawkins: *River out of Eden*, p. xiii.
26. André Pichot: op. cit., p. 769.

not muster the scientific elements essential to establish his theory."[27] The broadsides come not only from French territory, the Anglo-Saxons too can be stridently critical. Robert Koons writes: "Darwin's so-called theory is not really a theory at all: it is a schema for future theories," and he sees Darwin's voluminous writings as "a stack of promissory notes for future theories."[28] David Berlinski is still more cutting: "By the standards of the serious sciences, Darwin's theory of evolution remains little more than a collection of anecdotal remarks."[29]

Darwin was under attack from the very day the *Origin* was published. Even his "four musketeers" had issues of strong disagreement with him. Huxley, his 'bulldog', never believed in the process of natural selection and warned Darwin against gradualism. Lyell and Gray stuck to their opinion that variations (the "minor individual differences") were directed by providential influence rather than by 'scientific' chance. And Herbert Spencer "eagerly devoured Lamarck's evolutionary theories and, interestingly enough, he never really wavered from them, remaining unconvinced by Darwin."[30] Even Ernst Haeckel, the German scientist who was Darwin's most active propagandist, remained heavily influenced by Lamarck.

Of course, the objections from the Christian side never diminished in number or clamour. Copernicus' theory had undermined the Aristotelian universe of the Middle Ages and caused it to collapse, resulting in a general sense of insecurity and doubt. And see, now this Charles Darwin dared to assert in print that the human being, made in the image of God and therefore most noble among creatures, descended from the ape. An examination of the reactions to Darwin's thesis shows that, intuitively, intelligent people knew that there must be truth in the assertion, for the resemblances between human and primate are obvious. But every important new idea evokes the full spectrum of reactions, from enthusiastic approval over hesitation or bewilderment to frantic denial. And so did Darwin's "dangerous idea."

But: "Darwin may not have convinced people of natural selection. He did, however, convince them of the fact of evolution."[31]

---

27. Claude Allègre: *Dieu face à la science,* p. 132.
28. Robert Koons: *Why Darwinism Fails to Inspire Confidence,* pp. 21-22, in: *Uncommon Dissent,* William Dembski ed.
29. David Berlinski: *The Deniable Darwin,* p. 296, in: *Uncommon Dissent,* William Dembski ed.
30. A.N. Wilson: op. cit., p. 205.
31. Michael Ruse: *Darwinism and its Discontents,* p. 27.

What Darwin Could Not Know

In the standard version of history, the dawning of human culture happened about 10,000 years ago. Humans, mental beings, have always been asking the elementary questions about the beginnings of the universe, their own origin, the meaning of life and death. And they have always found or accepted answers to the questions of their ignorance. In the course of time the confined worlds of the human tribes have expanded, and the boundaries of their narrow egos have been relaxed, step by insecure step. Ignorance is fundamental; increasing enlightenment is the goal of humanity's crusade, still far from the Jerusalem that has to be conquered.

This sort of reflection is more than hyperbole. Science has always, in any period of history, assumed that it had reached the top of the hill of knowledge, striking up hymns of triumph. But looking back into the records of its errors, feuds and blind-alleys, one learns that even a Newton and an Einstein could be mistaken – actually could not but be mistaken – and that science, by way of speaking, keeps hitting its head against the wall of Truth or Reality. What, then, about our world of cars, airplanes and computers, are they not proof of the triumph of science? Actually it is a world for the most part created by technology, by the engineer. Science strives for understanding and knowledge, technology finds ways to make things – and existed long before our successful modern science. That both meet at many intersections should not efface the difference.

"If anyone was chasing a phantom or retreating from empiricism, it was surely Darwin, who himself freely admitted that he had absolutely no hard empirical evidence that any of the major evolutionary transformations he proposed had ever actually happened," writes Michael Denton in his *Evolution – A Theory in Crisis*. "There can be no question that Darwin had nothing like sufficient evidence to establish his theory of evolution. ... None of his claims received any direct experimental support until nearly a century had elapsed."[32] Michael Behe, another critic of Darwin, points out "how little Darwinian processes explain and how much is not understood."[33]

The fact is that Darwin *could not* substantiate his theory scientifically: the biological sciences in his day were not sufficiently advanced to provide the supporting evidence. He himself conceded: "It deserves

---

32. Michael Denton: op. cit., p. 117, 69, 79.
33. Michael Behe: *Darwin's Black Box,* p. 166.

especial notice that the more important objections [to his theory] relate to questions on which we are confessedly ignorant; nor do we know how ignorant we are."[34]

At the root of the origin of species lay the origin of life on our planet. Living things had to appear in non-living matter before they could branch out in a 'tree of life'. Today the extremely improbable accident of life's origin is as mysterious as ever. Darwin took the safe road and avoided the question. He wrote to a correspondent: "You expressed correctly my views where you said that I had intentionally left the question of the Origin of Life uncanvassed as being altogether *ultra vires* [beyond our powers] in the present state of our knowledge."[35]

We have seen how problematic the age of the Earth was in Darwin's time. Buffon, Hutton and Lyell had gradually increased the numbers from thousands to millions and hundreds of millions of years. "In 1795, James Hutton, the founder of modern geology believed that the surface features of the Earth were shaped gradually by incremental changes extending over enormous lengths of time. He realized that millions of years would be needed to accumulate rock sediments and to raise and erode mountains."[36] The huge time span of Lyell's geological eras was one of the reasons Darwin leaned so heavily on his gradualism. As he had no idea how the mechanisms of individual variation and inheritance could function, he had to posit minute gradual changes "over time", much, much time, to make his miracles acceptable.

Fossils are embedded in geological sediments which form strata. These findings were one of Darwin's two pillars of the evolution of the species (the other was the results of breeding, proof to him that "descent with modification" was possible). As Darwin did not know how old the Earth was, he could not know how old the fossils were. "The best he and other scientists of his day could say was that a given fossil came from a certain geological period."[37] At that time there were still enormous gaps in the fossil record. "During the 1850s, orang-utans, chimpanzees and gorillas were beginning to emerge from their jungle obscurity, and Richard Owen [one of Darwin's adversaries] dissected their bodies and studied their skeletons."[38]

The following paragraph from a science magazine dated March 1859

---

34. Charles Darwin: *The Origin of Species*, p. 447.
35. Paul Davies: *The Origin of Life*, p. 46.
36. id., p. 134.
37. Carl Zimmer: *Evolution*, p. 71.
38. id., p. 62.

gives a glimpse of the situation in Darwin's days: "In Africa there is a tribe of huge monkeys known by the name of Gorillas. Their existence has been known to white men for some years, but none have ever been taken alive. They live in the lonely retired seclusions of the forests, and the males are capable of coping in fight with the lion. The skull of one is in the Boston Museum, sent thither by the Rev. Mr. Wilson, a missionary. Last year, the body of one was sent from Sierra Leone to Prof. Owen, packed in a cask of rum. The males have a horrible appearance; they attain to a stature of five feet, with wrists four times the size of a man's."[39]

As far as Darwin knew, fossil evidence for human evolution had not yet been found. "Java Man" was discovered in 1891 by Eugen Dubois, who was unjustly ridiculed for suggesting that he had found a link between the primates and *Homo sapiens*. The "Taung Child", an *Australopithecus africanus*, was identified by Raymond Dart in 1924. Before Dart's discovery Asia, not Africa, was considered to be the cradle of mankind. "Peking Man", a *Homo erectus*, was found in 1926. And it was only in 1931 that the great discoveries started in central Africa. The problem of the missing link(s) will remain a bone of contention up to the present day.

The appearance of new species requires the inheritance of the useful variations. The lack of a mechanism for heredity was perhaps the biggest gap in Darwin's theory. "Darwin had been one of a long series of biologists who could make no real headway in understanding heredity. Indeed, Darwin's own ideas on the subject were far off the mark." (Eldredge[40]) "Darwin himself failed to solve the problem of speciation. ... In 1859, when he published the *Origin*, he actually did not have a single clear-cut piece of evidence for the existence of selection." (Mayr[41]) "Darwin was not able to present a single instance of speciation by natural selection in *The Origin of Species* ... He had to admit, when challenged, that he could provide no cases of animal-breeders producing a new species. Such breeding had definitely produced different varieties, but not a single new species." (Dennett[42])

It bears repetition: Charles Darwin could not know any of this indispensable data at the time of constructing his theory, nor could anybody else, simply because this knowledge was not yet available. "Darwin was

39. *Scientific American India*, March 2009, p. 8.
40. Denyse O'Leary: *By Design or by Chance?* p. 85.
41. Ernst Mayr: op. cit., pp. 193, 134.
42. Daniel Dennett: op. cit., p. 101.

acutely aware that the whole edifice he had constructed in the *Origin* was entirely theoretical. ... The highly speculative nature of his evolutionary model was quite apparent to Darwin himself ...'' (Denton[43])

Then what was the validity of Darwin's theory of evolution – of that what Darwin really did say? His system was a mental construction, an elaborated guess assembling and giving shape to various important ideas which were in the air. It was not based on scientific facts, but metaphorically on a geological (Lyell) and a socio-economic theory (Malthus).

A contention of this caliber demands expert support. Robert Koons writes: "Darwin's so-called theory is not really a theory at all: it is a schema for future theories. ... With the publication of *The Origin of Species*, Darwin produced a stack of promisory notes for future theories."[44] And David Berlinski: "By the standards of the serious sciences, Darwin's theory of evolution remains little more than a collection of anecdotal remarks."[45] Opinions like these are contradicted in most cases not on the grounds of Darwin's original system, but on what later has been put together as so-called 'Darwinism'.

Both Koons and Berlinski are outspoken critics of Darwin, as is André Pichot who terms the awe in which Darwin and his work are held "the Darwinian mythology." Daniel Dennett, on the contrary, is another 'arch-Darwinian', but he too has to admit that "no one knew better than Darwin himself the importance of anchoring a revolutionary theory in the bedrock of empirical fact, and he knew that he could only speculate, with scant hope in his own day of getting any substantive feedback."[46]

Darwin's Legacy

Darwin's Darwinism has met with difficult times in the course of its career and was around 1900 even declared death. But through gradual changes, additions and reinterpretations 'Darwinism' was "invented by the likes of Spencer, Haeckel, Galton, Weismann, de Vries. They were intentionally promoting Darwin, making evolution not only the central question in biology, but also of society, morals, religion, politics, etc.

---

43. Michael Denton: op. cit., pp. 55, 65.
44. Robert Koons: op. cit., pp. 21-22.
45. David Berlinski: op. cit., p. 296.
46. Daniel Dennett: op. cit., p. 149.

In short, they were making it into "the question that has remained fashionable."[47] That they have succeeded is illustrated by the fact that Charles Darwin came fourth in the BBC's 2002 "Great Britons" contest (behind Diana, Princess of Wales).

It is amazing how reverential materialistic scientists can be within the discipline to which they have dedicated their lives. Some evolutionary biologists have elevated Darwin above the angels. "At the Darwin Centennial in 1959, Julian Huxley delivered his famous oration announcing that evolution would be the new religion, leaving no doubt at all in the minds of those who were paying attention that many Darwinists were not simply disinterested scientists; they were definitely presenting a new worldview."[48] And David Berlinski writes sarcastically: "Daniel Dennett's assertion that natural selection has been demonstrated 'beyond all reasonable doubt' must be judged for what it is: the ecclesiastical bull of a most peculiar church, a cousin in kind to ecclesiastical bluff. When Steven Pinker affirms that 'natural selection is the only explanation we have of how complex life can evolve,' he is very much in the inadvertent position of the apostles. Much against his will, he is bearing witness."[49]

In the writings of evolutionary biologists one finds many references to passages in Darwin to prove a point. This, of course, is normal procedure. Yet the problem lies in the fact that Darwin, often in two minds, can be used to support all shades of opinion, even irreconcilable ones, like his own theory of evolution based on chance variations and Lamarck's inheritance of acquired characteristics. As mentioned before, this mixture of opinions and principles has been added to in every edition of the *Origin* Darwin published during his lifetime. He admitted: "The more I think the more bewildered I become."

The anti-religious evolutionists use Darwin as the epitome of materialism and atheism. This is evident in the mordant attack several proponents of 'Darwinian' evolution have recently been waging on religion, Richard Dawkins in *The God Delusion* and Daniel Dennett in *Breaking the Spell* among others. They rightly point to Darwin as their inspiration. Apparently hesitant and "agnostic," it doubtlessly was Darwin's intention to found his theory on scientific materialism, against the natural theology of clergyman Paley. As Stephen Gould put it: "Darwin applied a consistent philosophy of materialism in his interpretation of nature. Matter is the ground of all existence; mind,

---

47. André Pichot: *Histoire de la notion de gène,* p. 269.
48. Denyse O'Leary: op. cit., p. 153.
49. David Berlinski: *The Devil's Delusion,* p. 196.

spirit, and God as well, are just words that express the wondrous results of neuronal complexity."[50] Marcel Schutzenberger confirms this: "It was the intention of Charles Darwin to reduce the human being to an animal and to the laws of physics and chemistry."[51]

It is intriguing to find in the writings of practically every materialistic biologist a kind of sick pleasure to demonstrate how much their science abases the human being. The 'lord of creation' has become a naked or neotenous ape, a bipedal brachiator. Darwin set the tone: "In *The Expression of the Emotion in Man and Animals* (1872), Darwin undermined the traditional anthropocentric interpretation that divided animals from human beings. He destroyed the notion that God created the Earth and all of its creatures for humankind. As part of his evolutionary theory, he sought to show the continuity of species and that 'humans are not a separate divinely created species'."[52] According to Darwin: "Human beings are accidental and incidental products of the material development of the universe, almost wholly irrelevant and readily ignored in any general description of its functioning."[53] "We evolved to survive and reproduce, to get food and shelter and mates and to raise our children. We did not evolve to get insights into life's ultimate realities. What we do find is a bonus, and we expect that bonus to run out at some point."[54]

It was the debasement of the 'lord of creation' which contributed to Darwin's feeling that exposing his ideas in public would be like confessing a murder. But here again he is ambivalent. As John Carey remarks: "Darwin regards the 'victorious' forms of life as 'higher' than the 'beaten' forms, and at times describes the battle for survival in nationalistic and imperialist terms,"[55] congratulating himself to be an Englishman. Dennett even quotes him as having written "that man in the distant future will be a far more perfect creature than he now is."[56]

The dark side of science is the logical consequence of its materialism. Particles of matter, whatever their size, shape or activity, cannot miraculously produce something else than matter, for the basic premise says that there is nothing but matter. Matter cannot produce life or a conscious mind. May one, then, suppose that the materialist scientist

---

50. John Forster e.a.: *Critique of Intelligent Design,* p. 29.
51. Jean Staune and Eric de Romain: *L'Homme face à la science,* p. 170.
52. John Forster e.a.: op. cit., 124.
53. Scott Atran, in *What is Your Dangerous Idea,* p. 173, John Brockman ed.
54. Michael Ruse: op. cit., p. 290.
55. John Carey (ed.): *The Faber Book of Science,* p. 118.
56. Daniel Dennett: *Breaking the Spell,* p. 278.

suffers from a split brain, one part of which sees nothing than matter, mechanisms and machines in his laboratory, and another with which he loves his wife, adores his children, reasons, and practices his science?

Darwin could not prove any essential point of his theory, but behind it all was the intent to put it on a material basis, and to let all traditional, non-materialistic ideas loose as so many balloons children play with. The science of biology, in its development far behind to the exact mathematical and physical sciences, tried to catch up with them. It had to wrestle with a serious problem, for as the word 'biology' says, it is about the knowledge of life. Matter and life seem to be two different categories. It is not that difficult to get a grip on the material component of living things, but the something that makes them alive remains impalpable.

If the individual is a human animal, his society must be an organization of human animals. Its chief law will be "the law of the fittest", and the result must be *bellum omnium versus omnes,* a war of all against all. This interpretation of the human social relations, initially by Herbert Spencer, became known as 'social biology'; it would be the norm in the capitalist society as well as in international politics. If the fittest are the ones that survive, gain the upper hand and grow into superior beings, then the superior society or nation would be the one in which the fittest are intentionally, scientifically cultivated.

Closely related to this dominant view was "eugenics," usually associated with Hitler and Nazism, but widespread before the Second World War, especially in the Unites States. (See e.g. Edwin Black: *War against the Weak – Eugenics and America's Campaign to Create a Master Race.*) Darwin stated the principle himself: "With savages, the weak in body or mind are soon eliminated, and those that survive commonly exhibit a vigorous state of health. We civilized men, on the other hand, do our utmost to check the process of elimination; we build asylums for the imbecile, the maimed, and the sick; we institute poor laws; and our medical men exert their utmost skill to save the life of every one to the last moment ... Thus the weak members of civilized societies propagate their kind. No one who has attended to the breeding of domestic animals will doubt that this must be highly injurious to the race of man."[57]

If the human being consists exclusively of matter, then the way it functions is by mechanisms which make him into a Cartesian machine, or La Mettrie's *homme machine.* The former still allowed for a rational

---

57. Charles Darwin: *The Descent of Man,* p. 159.

soul; the latter drew the logical conclusion and denied the existence of a non-material soul too. Neurobiology as practiced nowadays reasons from precisely the same standpoint: the mind is a formation of the body, thought a secretion of the brain. Under reference to Darwin, any non-material phenomenon is automatically discarded as paranormal, supranormal, metaphysical or, worst of all, mystic. The brain is a product of material evolution, adapted to the increasing complexity of the evolving species. Evolutionary psychology now teaches us that our brain developed in the Stone Age populations of "hunter-gatherers," and that we are forced to obey its primitive workings.

Darwin's thought was only indirectly derived from the Enlightenment and belonged to the coarser post-Enlightenment of the nineteenth century. Eighteenth-century Enlightenment, the Age of Reason, was quite idealistic and put Humanity on a pedestal; it also had an absolute faith in progress. Because of the industrial revolution, the rising tide of the masses and the commercial spirit of the bourgeoisie, the idea of progress had been tied to the ideal of science. But if matter was the one and only substance, and the human animal had evolved from the primate, what kind of progress, if any, could be expected in the long term?

Moreover, if the evolution of the species was a material, mechanical process, was it not a misinterpretation to qualify what evolved later as necessarily being higher? Higher according to which scale? Materialism had collapsed all existential levels of the Chain of Being and of Linnaeus' classification into one: matter. That the human being thought of itself as higher than all other beings was vainglorious fancy. Copernicus had put the Earth in its right place as a planet among the other planets; Darwin had put the humans in their right place as animals among the other animals. Most Darwinian authors enjoy reminding their human readers of this "fact," rubbing them with their noses into their animality, so to speak.

Finally, Darwinism is so often used as an argument against anything related to God, religion, spirituality, occultism or the paranormal, that it has become synonymous with godlessness. As mentioned before, some neo-Darwinians have gone on an all-out attack on creationism and "intelligent design," a movement that in its scientific form finds the works of nature too complex to have been brought about by the material mechanism of 'Darwinism'. We know about Darwin's religious struggle, which resulted in agnosticism. His present-day followers are probably right in supposing that their idol, at bottom, was a religious non-believer. If so, however, the cause was much more the influence of the general intellectual attitude of his century on him than it was his own decision. Darwin always had difficulty to arrive at a decision. That he had nonetheless the courage to take some very important decisions, realizing the import of their ineluctable impact, is a main characteristic of his stature.

4.

Lamarck: The First Evolutionary Theorist

Who would dare to put boundaries to the human intelligence, and to assure that there is a knowledge man will never acquire or a secret he will never penetrate?

JEAN-BAPTISTE LAMARCK

The Natural Sciences

Members of the French nobility had names as long as a freight train – so too Jean-Baptiste-Pierre-Antoine de Monet de Lamarck, usually referred to as Jean-Baptiste Lamarck, or for short Lamarck (1744-1829). The fame of this extraordinary scientist has been eclipsed by the aura enveloping Charles Darwin. When Lamarck is mentioned, it is mostly in "a systematically denigrating way, often ill-founded, not only by numerous biologists but also by historians of biology."[1] This is without exaggeration one of the great injustices in the history of science. Lamarck's vitally important work, which would make Darwin possible, is suppressed or misrepresented because of a lack of knowledge; and the scientist himself is condemned to oblivion, even though his stature equalled Darwin's.

Lamarck, after a brief career in the military was cut short because of an injury, entered science as a botanist. With the support of his mentor in the early period of his life, the great Buffon, he became the curator of the Parisian Botanical Garden. With his thirst for knowledge and his encyclopedic mind Lamarck published before long the basic reference work in French botany, *La Flore française*, consisting of three fat volumes. It made him at once the most famous botanist in the land.

In 1793 his career took another sudden turn: he was offered the chair of "insects and worms" at the newly founded National Museum of

---

1. André Pichot: *Histoire de la notion de vie*, p. 686.

Natural History in Paris. This was an academic position of low esteem, supposedly presented to him because nobody else was interested. All the same, it provided Lamarck with a stable and well-remunerated job, and he dedicated himself to this new field of study with his considerable capacities and innate zest.

The invertebrate animals, to which the insects and worms belong, were of course included in Linnaeus' general classification, but they had never retained the Swede's attention. Their classification was still chaotic and the knowledge about them rudimentary, not to say often fanciful. Lamarck would change all that and become the real founder of the classification of the invertebrates (the animals without backbones), which, for the first time, he clearly separated from the vertebrate animals (the animals with backbones). He himself would describe no less than one thousand species of invertebrate fossils. The volumes of his classical *Natural History of the Invertebrate animals,* showing "a knowledge *sans pareil,*" was published from 1815 to 1822.

Lamarck would in fact be the first to sketch a scientifically warranted 'tree of life', some fifty years before Darwin. The separation of the invertebrate animals from the vertebrate ones showed the hierarchical structure in the evolutionary descent of the latter. Attention may once more be drawn to the direct influence of Linnaeus' classification on the mind of a naturalist like Lamarck and thus on evolution, and to the silent presence of the hierarchical structure of the chain of being behind it all.

Founding Biology

"Lamarck is truly the inventor of biology as a science of life and of the living beings. Not only did he invent the word 'biology', he was also the first to understand biology as an autonomous science – a science not only distinct from physics and chemistry, but also from taxonomy [the classification of living beings], anatomy, physiology and medicine. The subject of this science was to study the characteristics common to plants and animals by which they are distinguished from inanimate bodies. This new science had to study the living beings as being alive and therefore different from inanimate objects." Thus writes the French historian of science André Pichot.[2]

---

2. André Pichot: *Histoire de la notion de vie,* p. 588.

Lamarck used the word 'biology' for the first time in 1802. It is formed from the Greek word 'bios', meaning 'life', and the commonly used '-logy', from 'logos', meaning 'word' or 'knowledge'. In one of his writings he defined the term as follows: "Biology comprehends everything that has to do with living bodies, and particularly with their organization, their developments, with their increasing complexity and the movements of life, with their tendency to create specialized organs, to isolate them, to centralize their action around one point, etc."[3] At first sight this definition looks quite abstract, but each term has its concrete significance in Lamarck's way of thinking, as we shall see.

On top of all that, Lamarck was one of the founders of paleontology, the study of living beings in times very long ago based on remains of their bodies. The questions about life in those times presented themselves to his discerning mind while looking for the petrified relics of ancient life, which he called 'fossils' – a word also coined by him. Some ideas about fossils, even at the end of the Enlightenment, were still rather odd. Biblical literalists continued thinking that God had not only created the animals, but the fossils too. Fantasies about the gigantic, apparently misshapen or strangely familiar bones were rife. Lamarck was one of the first to see their fundamental importance within the framework of an evolutionary schema.

He found, for example, an explanation for the fact that fossil shells of *Nautilus pompilius* were discovered not only in places in France, but also in "the ocean of Greater India and the Moluccas", the Malay Archipelago; from this he deduced that at one time the landmass of which France was a part must have been under the sea. (Leonardo da Vinci had already drawn the conclusion that mountains must have been covered by the sea when fossil sea shells were brought to him from the Dolomites, but two centuries ago little was known about the scientific knowledge of the Florentine.) Lamarck also noted that the same fossils were found in the same sedimentary strata of France and Great Britain. He concluded correctly that both land masses had been connected in a far past, but that they had been separated afterwards by changes in the sea level. Systematizing this branch of his newly gathered knowledge, Lamarck founded what he called the science of 'hydrogeology'.

Confronted with all these new data, the importance of time, much time, became vital. The sciences of the Earth and of life demanded an "immensity" of geological and paleontological time. "Oh, how

---

3. Goulven Laurent: *La naissance du transformisme,* p. 110.

enormous is the antiquity of the terrestrial globe!" wrote Lamarck. "And how small are the ideas of those who attribute to the existence of this globe a duration of six thousand years from its origin to the present! And how much this antiquity of the terrestrial globe will still be increased in the eyes of man when he will have obtained a correct idea of the origin of the living bodies, as well as of the gradual causes of the development and the perfection of the organization of these bodies, and especially when he will realize that the time and the circumstances have been necessary to bring all these living species into existence as we see them at present, and that he himself is the last result and at present the maximum of that perfection, of which the end, if there is one, cannot be perceived!"[4]

Linnaeus had accepted the time scale as calculated by Archbishop Ussher, according to whom the Earth was 6000 years old. Buffon had prudently suggested an age for the Earth of 70 million years. For Lamarck, James Hutton and Charles Lyell the perspective backwards and forwards in time became endless, or at least indefinable – for only in this way could the strata of the Earth's crust and, in them, the fossils of paleontology be explained. A biographer of Charles Lyell recognizes that it was Lamarck who provided the instrument of stratigraphy and that he is therefore, with Cuvier, one of the founders of paleontology. "Lamarck's great systematic study of invertebrate zoology [and of 'hydrogeology'] had been the indispensable foundation for the study of invertebrate fossils and their use in the identification of strata,"[5] the layers of the Earth's crust formed during long time periods by sediments of fossils, plants and erosion of the soil.

The number of new sciences which Lamarck founded or systematized – and we have not even mentioned meteorology – may perplex the reader. This, however, was an axis time for what are now called the biological sciences. There had been Aristotle, Galen, Descartes and La Mettrie, Paracelsus, Vesalius in Padua and the naturalists connected in one way or another with the University of Leiden, that practically forgotten but important former centre of medical research;[6] each of those names, and many more, are worth a chapter if not a volume in the history of biology. But what is now known under this name was systematized mainly by Jean-Baptiste Lamarck. It is noteworthy that

---

4. Goulven Laurent: *La naissance du transformisme*, p. 87.
5. id., p. 65.
6. See e.g. Matthew Cobb: *The Egg & Sperm Race – The Seventeenth-Century Scientists Who Unravelled the Secrets of Sex, Life and Growth.*

this take-off of the biological sciences came more than a century after Isaac Newton had published his *Principia Mathematica* (1686). Physics, based on mathematics, was considered a true, 'hard' science; the 'soft' sciences of which biology was comprised tried desperately to catch up with physics and to base themselves also on the scientific method.

As Lamarck wrote: "For the man who dedicates himself to a career in the sciences, more particularly the physical sciences, and who wants to contribute to their progress, nothing is more advantageous than a strictness in the principles which allows him to deal only with exact knowledge, which makes him suspicious of any supposition or guess-work, and which makes him, above all, acquire the important habit never to confound what is truly proven with what is simply apparent. It is certain that for whomever wants to penetrate the secrets of nature, no quality is more desirable than the strictness I have just mentioned."[7]

Checking the dates and the events in Lamarck's life, it will be obvious that much of his scientific career was intertwined with the French Revolution and the Empire, the period in French history from 1789 to 1815. The Botanical Garden and the *Muséum* were creations of the Republic, and Lamarck's mental outlook was that of the Enlightenment which led to the ideals of the Revolution. On the one side fanatical Jacobins killed the chemist Antoine Lavoisier, on the other the Revolution promoted the careers of naturalists like Cuvier and Lamarck. Generally speaking, it was a great time for new ideas in science, as is proven by the numerous discoveries and innovations. (Napoleon took a platoon of scientists with him on his military expedition to Egypt.) But the historical context had also its negative consequences for the acceptance of that science, especially in Great Britain, and for the differences in the scientific approach between the Anglo-Saxon countries and France up to this day.

God or Nature

The existence of God, from unquestioned, became ever more problematic and was already denied by many atheists. In his writings Lamarck still professed to be a deist – "a vague deist" – in the way Voltaire had been: recognizing the existence of God as creator, but removing him outside the works of his creation. God was "the Supreme

---

7. Goulven Laurent: op. cit., p. 138.

Being" who had created nature with its laws, and who had in this way established an "order of things' according to which the world developed or evolved in the course of time. "It has been supposed that each species was invariable and as old as nature," wrote Lamarck, "and that it was especially created by the sublime Author of all things. But can we prescribe rules to him for the execution of his will and fix the manner he should follow in this? Can his omnipotence not have created an 'order of things' which successively brought into being everything we see, as well as everything that exists and that we do not know?"[8] Nature, having been created by the Supreme Being, became a kind of substitute of God, possessing the powers to modify and develop. "I hope to prove that nature possesses the necessary means and the faculties to produce herself what we are admiring in her," he wrote.[9] Nonetheless, materialists like Lamarck posited that nature and everything in it was material and nothing else.

Lamarck is often erroneously labelled as a vitalist. He saw himself as "a materialist through and through," although materialism then was not exactly what gross materialism is now. He wrote e.g. about the existence of the soul: "As to me, without repudiating any matter of religious belief or which can be a consolation to an honest person [*un homme de bien*] who has convinced himself of it, I say that this kind of consideration is absolutely foreign to my subject. For both the immortal soul of man and the perishable soul of the animals remain unknowable to me physically."[10] "Contrary to Buffon and many others, Lamarck declares that there is no living matter as such, nor is there a vital force. Matter is everywhere of the same nature, in the living beings as well as in the structures of the minerals. What is proper to life, is its *organization*. Living beings are *organized* beings." (Claude Allègre[11]) The quite pedestrian word 'organization' acquires here the unexplained, quasi magical power which others ascribe to 'life'.

But "life is a mechanism," a material mechanism, as is everything else. Lamarck, like Darwin, had gradually changed his outlook from that of a believer in 'fixism', a variant of creationism, to that of a materialistic evolutionist. The tradition initiated by Descartes was very strong, especially in France (*Descartes, c'est la France!*). According to René Descartes (1596-1650) the human being consisted of a body and a rational

---

8. André Pichot: op. cit., p. 588.
9. André Pichot: *Histoire de la notion de gène,* p. 266.
10. Goulven Laurent: op. cit., p. 105.
11. Claude Allègre: *Dieu face à la science,* p. 127 (emphasis added).

soul. His body, and that of all animals, was a mechanism, a machine; his soul was an "epiphenomenon" – one of those labelling terms which can mean everything without meaning anything. It will not take long before Julien de la Mettrie and his generation keep the mechanical body and discard the soul, on the same ground as the one of Lamarck quoted above, namely that "it remains unknowable physically."

Knowledge is a search which starts from a fundamental ignorance. The gains of knowledge are hard-won and always limited; ignorance remains the pond on which float the lotus leaves of the acquisitions of human knowledge. Being an action of the mind, knowledge tries in vain to get a firm grasp on that-what-is, on Reality. It forms mental projections which try to circumscribe, to encompass Reality. One such projection is mechanical toys. The human being has always been fascinated by automata – from Vaucenson's quacking duck in the 17th century to the futurist 'humanoid' robot – made to imitate the mystery of life. Mechanisms and automata were among the favourite playthings of the rich in the Renaissance (the engineer Leonardo was a master at designing and fabricating them), and afterwards at the courts of the European princes.

The comparison between living things and automata came naturally: "The picture of organisms that emerged from seventeenth-century science is filled with mechanical metaphors: the stomach as a retort, veins and arteries as hydraulic tubes, the heart as a pump, the viscera as sieves, lungs as bellows, muscles and bones as a system of cords, struts and pulleys ... Thus each organic body of a living thing is a kind of divine machine, an automaton fabricated by nature, which infinitely surpasses all artificial automata," but which the materialist supposes to function in the same way.[12]

It is this mechanistic simile, created by an ignorance seeking knowledge, which has become one of the basic ways of reasoning of scientific materialism. What the mind, embedded in materiality, cannot know is held to be nothing but fancy which, at long last, should be swept aside. Reality, as mapped by the mind, may only be approached from the outside. The more numerous the accumulated data, the more accurate the reproduction, the map of reality. But as long as that what is real can only be approached from the outside, the reality of the real cannot be known.

---

12. Niall Shanks: *God, the Devil and Darwin*, p. 32.

The Progressive Upward Climb of Life

Nowadays the thought will come to hardly anybody that evolution does not happen from below upwards, from the simple to the more complex, from the one-celled protozoa to the mammals, but in the opposite direction, from the complex to the elementarily simple. Yet, the latter was the way in which evolution was at first supposed to have happened: as a degeneration. The donkey might be a degeneration of the horse, the monkey of the human. Buffon (1707-88), for example, mooted the opinion that originally "among the animals and even among the plants there might be not several species, but only one, which brought forth the other species by degeneration. If it were true that the donkey was a degenerated horse, there would be no limits to the powers of Nature, and one would not be wrong to suppose that, from one single being, she was in the course of time able to bring forth all the other organized beings."[13]

When Lamarck put together his transformist theory in the years 1797-1800 the direction of the evolution was an as yet unsolved problem on which he had to take a decision. "If all the subdivisions which the animal kingdom comprises necessarily form a series of masses [i.e. populations or species] in the increasing or decreasing order of their composition, does one in the disposition of the series proceed from the more complex to the more simple, or from the more simple to the more complex?"[14] However odd it may now seem to us, this was quite a reasonable question, given the facts that the story of a direct and perfect creation had been ingrained in the Western minds for centuries, that progressive change from an amoeba to an elephant is far from obvious, and that the biological sciences stood only just on the threshold of a scientifically founded knowledge and classification. It was a fundamental question.

The Chain of Being and Linnaeus' classification had suggested the answer, and the ideas about evolution which were 'in the air' in the century preceding Lamarck pointed in the same direction: from the simple to the complex. But it was Lamarck's enormous erudition that allowed him to definitively direct the arrow of the evolutionary trend in the direction accepted to this day. In his words: "It is therefore of extreme importance for the furthering of our future knowledge that

13. Goulven Laurent: op. cit., p. 35.
14. Goulven Laurent: op. cit., p. 50.

we introduce in the general distribution of the animals a reversal, putting the most imperfect and simple animals at the beginning of the distribution [i.e. the branching off], while the more perfect, those with the most complex organization, will come at the end."[15] (Note the keyword "reversal.") In one of his works at the time, Lamarck "again states clearly that he has discovered the fundamental connection which exists between classification and descent." His system of the process of evolution as concretized in the history of the living beings gained instantaneous authority; it was e.g. copied as early as 1816 by a French naturalist as "Sketch of the descent of the animals from the infusorian to the monkey."

Another fundamental problem was the origin of life. How had life appeared on planet Earth in an entirely material environment? Those "infusorians" – Lamarck's word for the first life forms which he also called "monads," or "matter barely animated" – where had they come from, how had they come about? Lamarck had no idea, just like Charles Darwin will have no idea and we still have not the foggiest.[16] Therefore he accepted, like anybody else at the time and in previous times, the *generatio spontanea,* spontaneous generation. Life had appeared just like that, all by itself, automatically. (It is the inability of any scientific theory to answer a crucial question like this that makes many questioning people feel safe in a creationist worldview.)

However, this did not prevent Lamarck, as well as Darwin, to cover his ignorance with a screen of clever invention. The *génération directe,* as he called spontaneous generation, "consists simply of the arrangement, by nature, of a small gelatinous mass, and the creation within it of an interaction of fluids."[17] How did such a small gelatinous mass become a living being? Well, simply again, by the action of heat, electricity (a new fad at the time) and fluids (which may prefigure Darwin's "warm pond"); these three elements suffice as the "stimulating cause." – "Nature, among the bodies which have resulted from her operations, has been able to form some which could react to the first effects of an organization and the movements which constitute life. This is indeed what she seems to have done by producing, among the inorganic bodies, very small gelatinous bodies of the weakest possible consistency."[18] The composition of the cell was still unknown. But the knack of fabricating

---

15. id., p. 70.
16. See e.g. Robert Shapiro: *Origins.*
17. André Pichot: *Histoire de la notion de vie,* p. 637.
18. Goulven Laurent: op. cit., pp. 112-13.

explanations to cover gaps in the knowledge will continue to be a frequent feature in the biological sciences.

Synthesizing his exceptional knowledge from diverse fields of research, Lamarck announced his theory of "transformism" in the year 1800. The word "evolution" was at the time associated with other life processes, especially the development of the embryo, and will not even be found in the first edition of Darwin's *Origin*, where preference is given to the term "transmutationism." Goulven Laurent notes: "It should be marked that the notion of the transformation of the species, introduced by Lamarck into the scientific domain, has not gone away since then, and that it has remained the keystone of the whole evolutionary edifice."[19]

All great persons seem to have a nemesis. For Newton it was Leibniz, for Darwin it will be Richard Owen – and for Lamarck it was Georges Cuvier (1769-1832). Cuvier was one of the great French naturalists and throughout his life much honoured in his country, for in his photographs his chest is bedecked with medals and orders. He was the founder of vertebrate paleontology, and became legendary for being able to reconstruct an animal unknown to him from a single of its bones. And he was a "fixist," which means that he did not accept that life forms evolved over time; once originated, probably by creation, they remained the same forever.

Still, the fossils of the giant ground sloth, the Irish elk, the American mastodon, and many others, were proof that species had become extinct. Because he said so overtly, Cuvier was attacked by the Biblical literalists, "who could not believe that God, having created all things and pronounced them good, would allow any of them to be wiped out."[20] His reasoning led to the formulation of "catastrophism": gigantic catastrophes (the last one being the biblical Flood) explained the disappearance of so many species. From this standpoint he lashed out with his bemedalled authority against the gradualism of Lamarck on any possible occasion. Thanks to the prestige of Darwinism, gradualism won the day – for more than one and a half century, till the great biological extinctions had become incontrovertible, and Stephen Gould and Niles Eldredge formulated their theory of "punctuated equilibrium" in the 1970s, causing a revival of the battle between fixism and gradualism.

Linnaeus had classified *Homo sapiens* among the primates, but it was

19. id., p. 82.
20. http://www.ucmp.berkeley.edu/history/cuvier.html

Lamarck who showed, for the first time, that the human *descended* from the primate. Like all other living beings, the human was a product of nature, having acquired his particular complexity in a succession of generations through vast stretches of time and favourable circumstances. "Quadrupeds were transformed into bipeds." The text in which Lamarck wrote this about the human descent was the first which described the evolution from monkey into human. This development, like all other transformations, had come about through the mechanism of evolution. "And so the human being was inserted into the category of the animals."[21]

If everything is the product of nature, *is* nature, and therefore material, so must the human mind be too. "Let me ask what is that particular something that is called mind," wrote Lamarck. "One says that it is special, related to the activities of the brain in a way which renders the functions of this organ of a different order than those of the other organs of an individual. I can only see, in that fictitious something without any counterpart in nature, a means of the imagination to solve the difficulties which one has been unable to eliminate for lack of having sufficiently studied the laws of nature."[22] What is here translated as 'mind' is *esprit* in the French language, which can mean 'spirit' or 'mind' (and even 'wit'). In Lamarck the first significance was out of the question. From a spark of the Divine the spirit or soul had been degraded to a rational, albeit still immortal being. Now that the mind had become the epiphenomenon of a machine, it was not even taken into consideration anymore. All life was matter and mechanism.

Transformism

Lamarck first made his evolutionary theory public in the year 1800. The seminal book in which he disclosed his system in full, *Philosophie zoologique*, was published in 1809, two hundred years ago at the time of writing the present essay. 2009 is the two hundredth anniversary of Darwin's birth and the one hundred and fiftieth of the publication of *The Origin of Species by Means of Natural Selection*. Both events are being remembered by a spate of articles and special programs in the media, and of international meetings and commemorative conferences. In this

---

21. Goulven Laurent: op. cit., p. 117.
22. id., p. 118.

commotion, Jean-Baptiste Lamarck is routinely overlooked and could as well never have existed.

It matters to sketch his theory briefly, if only because it is so often ignorantly distorted even by well-known authors. The general basis of Lamarck's thesis is that in the course of time "transformation" (i.e. evolution) of the species has taken place, from the elementary simple "monads" of life to the most complex species of animals, with the human at the top of them. On this basis Lamarck states that within the species there is a tendency to increase the complexity of their organization, resulting in numerous differentiations and specializations. In this way the species become more "perfect", which is Lamarck's word for 'evolved'. According to him "it is the tendency of the living beings to increase their complexity which is the driving force of the evolution." (Pichot[23])

The tendency of increasing complexity can not be realized unhindered. It is checked by the external circumstances, by the environment, which forces the species to develop certain functional shapes suited for survival in their environment. The circumstances create the *besoin*, the inner need to develop new organs for the organisms to be able to thrive in those circumstances; the need creates habits; the habits create organs. Lamarck puts it as follows: "The circumstances have an impact on the form and the organization of the animals ... Great changes in the circumstances lead within the animals to great changes in their needs, and such changes in the needs lead necessarily to changes in their actions. Consequently, if the new needs become constant or quite durable, the animals adopt new habits, as durable as the needs from which they have originated."[24]

His second principle was: *la fonction fait l'organe*, sometimes rendered in English as: the use creates the organ, and consequently non-use makes an organ shrink and eventually even disappear. The word *fonction*, in this context, is ambivalent. If an organ does not yet exist, it cannot 'function'; therefore it might be more accurate to understand that it is the habit, expression of a need, which creates the organ. The classical example of this evolutionary phenomenon is the neck of the giraffe. The need to find food higher-up in the trees than where other animals could attain, created the habit to reach for it and consequently the long neck. (Darwinists seldom omit to quote with condescending

---

23. André Pichot: op. cit., p. 658.
24. André Pichot: op. cit., p. 667.

irony this example of Lamarckian naivety, unaware of the fact that Charles Darwin himself used the same example, which he may have borrowed from Lamarck.)

Lamarck called the modification of the organs, because of the changes in the needs and the habits necessitated by the environment, his first law. He formulated his second law thus: "All what nature has caused the individual [plant or organism] to gain or to loose because of the influence of the circumstances to which its species has been exposed for a long time, and consequently because of the influence of the predominant use of a certain organ, or because of an enduring defect of a certain part – all this she passes on by means of heredity to the new individuals which result from it, on condition that the acquired changes are common to both sexes or to those which have produces those new individuals."[25] This is the famous Lamarckian law of "the inheritance of acquired characteristics."

That characteristics which organisms had acquired during their lifetime were passed on to their offspring was generally accepted from Aristotle till after the publication of *The Origin of Species*. (It is still a widespread popular belief, as one can easily find out from expectant mothers and in discussions of family traits. Didn't Darwin fear that his children would inherit his physical deficiencies?) The amazing fact is that Lamarck himself never came up with a theory explaining the inheritance of characteristics acquired in the Lamarckian way. He had no scientific explanation of the way heredity worked and he never proposed one. "This absence of a theory of heredity may surprise us, considering the 'transformist' ideas of Lamarck and the part played in them by the transmission of the character traits. It is the weak point in his thesis. No doubt, he accepted the notions of heredity which were common in his time." (Pichot[26])

Yet, who *did* propose an explanation of heredity was Charles Darwin! He had not the faintest scientific proof of how heredity worked, but his mind was as fecund as that of any biologist in fabricating explanations where they were needed but lacking. Darwin's explanation of heredity was his theory of "pangenesis" and the "gemmules." The cells in all parts of the body of a living being secreted small particles containing duplicates of their essence. These particles were absorbed by the bloodstream, which carried them to the sex organs, where they

---

25. id., p. 679.
26. André Pichot: op. cit., p. 641.

concentrated in the eggs and in the sperm, and were thus transmitted to the offspring. Present-day Darwinians often cover this fanciful theory with the veil of discretion. But it is from the word 'pangenesis' that the word 'gene' has been deduced.

In brief, the evolutionary principle of the inheritance of acquired characteristics, for which Lamarck is almost exclusively famous, was at the time the current opinion, also among the naturalists, and it was turned into a theory not by Lamarck but by Charles Darwin.

Caricature

Then why, among the theoretical explanations of the fact of evolution, seems the field chiefly divided between two camps: a majority of Darwinians and a minority of Lamarckians? According to Lamarck the factors which cause the transformation of species are the external determinants of the environment. They create in response a *besoin*, an inner need, which ultimately creates modifications in the physical bodies. According to Darwin, on the other hand, the modifications in the species are caused by the internal factor of random variations, now called mutations, among which the favourable ones are selected in the confrontation of the organisms with the environment (the process of adaptation). Both views, as Pichot remarks, could be merged into one, but for the moment they are still stubbornly confrontational.

Although 'Darwinism' gained the upper hand, Lamarckism has always remained present in the background. It increases in importance every time the impact of the environment comes into play; it is now broadly accepted in the cultural field of evolution; and it is returning to full stature because of the recent discoveries in epigenetics. Then why does one continue reading statements like: "Lamarckism is false – in the biological world there is no inheritance of acquired characteristics" (Michael Ruse[27]); or: "Lamarckian evolution was downright spiritual" (Robert Wright[28]) when Lamarck was a declared mechanistic materialist; or: "Thanks to modern genetics, we know that Lamarck's theory of acquired characteristics cannot happen" (Michael Shermer[29])?

The 'Darwinian' camp, as usual, refers its opinions to their source

---

27. Michael Ruse: *Darwinism and its Discontents*, p. 262.
28. Robert Wright: *The Moral Animal*, p. 232.
29. Michael Shermer: *In Darwin's Shadow*, p. 117.

in holy scripture, Darwin's texts. Darwin wrote disparagingly about Lamarck as he did about his grandfather Erasmus Darwin, although the pioneering thought of both of them had provided essential elements of his hypothesis. The recluse of Down may not have been an unfair man, but he could become extremely possessive where his life's work, his theory of transmutationism with its irksome problems, was at stake. In this he did not differ from Newton or, in more recent years, Kelvin or Eddington, and many others.

Nevertheless, "the attitude of Darwin and the Darwinians towards Lamarck bestows honour neither on their intelligence nor on their intellectual honesty," writes Pichot,[30] who pillories the systematic and often ill-founded denigration of which Lamarck is the object with many biologists and historians of biology. We will see that some quite level-headed authors allege that 'Darwinism' has become a fanatical sect, a scientific Church. Darwin, like Comte, Freud and Jung, did indeed show traits of the founder of a faith, while remaining subject to devouring doubts. The anti-Lamarckism and the false rumours it has turned into clichés seems to be the result, at heart, of a metaphysical faith in materialism, or a faith in metaphysical materialism, traceable to Darwin himself.

From this one should not conclude that (true) Lamarckism is no longer a factor in evolutionary thought and research. Burton Guttman writes in his book *Evolution*: "Lamarck's ideas are worth mentioning if only because similar ideas, labeled 'neo-Lamarckism', keep reappearing in biology." And Stephen Gould wrote: "Cultural change manifestly operates on the radically different substrate of Lamarckian inheritance, or the passage of acquired characteristics to subsequent generations. [Even a well-read biologist like the late Gould still identified Lamarck with 'acquired characteristics'.] Evolutionists have long understood that Darwinism cannot operate effectively in systems of Lamarckian inheritance, for Lamarckian change has such a clear direction and permits evolution to proceed so rapidly that Darwin's much slower process of material selection shrinks to insignificance before the Lamarckian juggernaut."[31] One could go on quoting in the same vain and show that even Francis Crick, on alternate days, dares to lean towards Lamarckism.

The following words of André Pichot are worth quoting in conclusion of this chapter: "Lamarck's thought, though often neglected, is a

---

30. André Pichot: op. cit., p. 941.
31. Stephen Jay Gould: "More Things in Heaven and Earth", in Hilary and Steven Rose: *Alas Poor Darwin,* p. 98.

monument in the history of biology, as much as the thought of Aristotle, unquestionably the only one with which he can be compared."[32]

32. André Pichot: op. cit., p. 686.

5.

Alfred Wallace: The Other Darwin

My contribution is made as a man of science, as a naturalist, as a man who studies his surroundings to see where he is. And the conclusion I reach is this: that everywhere, not here and there, but everywhere, and in the very smallest operations of nature to which human observation has penetrated, there is Purpose and a continual Guidance and Control.

ALFRED WALLACE

The Amazon and the Malay Archipelago

In age Alfred Russel Wallace could have been the younger brother of Charles Darwin, for Charles was born in 1809 and Alfred only fourteen years later, in 1823. In social status, however, the difference was considerable. Charles belonged to the rich upper class, was educated at top institutions in Britain, never had so much as a hint of financial problems and could, because of his standing, rely on the support of people who counted. Alfred, on the contrary, did not even finish grammar school, was all his life troubled for money, and belonged to the grey crowd of the unknown. "Degreeless, and without an important institute affiliation, Wallace nevertheless carved a path through life," writes Michael Shermer.[1] Wallace himself encapsulated the inner drive which carried him through it all in one sentence: "I possessed a strong desire to know the causes of things."[2]

One of the jobs which he took up in his youth, thanks to his

---

1. Michael Shermer: *In Darwin's Shadow – The Life and Science of Alfred Russel Wallace,* p. 42. Shermer is founder and author of *Skeptic Magazine,* and known as a debunker of phenomena supposedly outside the bounds of scientific materialism. His excellent biographical study of Wallace is nevertheless evenly balanced. The present author is heavily indebted to it.
2. id., p. 48.

intelligence and studious reading, was that of surveyor. It was the initial boom time for building railway lines. Stephenson's steam locomotive, the *Rocket,* had been on the rails since 1830, and railway tracks began to crisscross through the landscape, of course first in Britain, but soon afterwards also in other countries. The railway fever caught on despite dire predictions that steam locomotives would start fires along the tracks, that the cows would become hysterical and no longer produce milk, and that the high speed (up to 40 km per hour!) would cause dangerous physical symptoms in the passengers. Surveying included map making and much time to spend in the open country. The skill needed for map making provided Wallace with a temporary job as a teacher, the time spent in the open gave him the occasion to indulge in his passion for beetles and all kinds of creepy crawlies.

He studied for some time in schools called "Mechanics' Institute," intended to give a basic education to youths of the working class. He also read everything he could dig up in the libraries to which he had access. The level of his interests can be deduced from some of the titles he would refer to in later years: Darwin's just published account of his voyage on the *Beagle,* Thomas Malthus' *Essay on the Principle of Population,* and Alexander von Humboldt's *Personal Narrative of Travels to the Equinoctial Regions of the New Continent,* "the first book that gave me the desire to visit the tropics." The two last works played an important role in Darwin's life too, and there must have been many more books which contributed to the formation of their doctrines in both men.

Despite the differences in social standing, the general intellectual background of the age was obviously the same for both of them. Wallace too was influenced by the ever increasing secularization. The disappearance of the concept of the Christian God, and "the final removal of the Deity from nature in the nineteenth century" affected his worldview as well. He became an out-and-out freethinker. "By the time I came of age I was absolutely non-religious. I cared and thought nothing about it, and could be best described by the modern term 'agnostic'" – the same term Darwin applied to himself. At the age of thirty-eight Wallace will write in a letter: "I remain an utter disbeliever in almost all you consider the most sacred truths. ... I am thankful I can see much to admire in all religions ... But whether there be a God and whatever be His nature; whether we have an immortal soul or not, or whatever may be our state after death, *I can have no fear of having to suffer for the study of nature and the search for truth,* or believe that those will be better off in a future state who have lived in the belief of doctrines inculcated from childhood, and which are to them rather a matter of blind faith than intelligent conviction."[3]

A book that influenced Wallace particularly was Robert Chambers' *Vestiges of the Natural History,* published anonymously in 1844. "*Vestiges* started harmlessly enough," describing the solar system and the universe as it was conceived at the time. Then "Chambers worked through the geological record, noting the rise of fossils throughout history. The simple appeared first and then the complex. As time went by, higher and higher forms of life left their mark. And then Chambers made a scandalizing claim: if people could accept [since Newton] that God assembled the heavenly bodies by natural laws, 'what is to hinder our supposing that the organic creation is also a result of natural laws, which are in like manner expressed by his will?' That would make more sense than God stepping in to create every species of shrimp or skunk." (Carl Zimmer[4]) *Vestiges* was one of the first texts openly advocating a naturalistic interpretation of the living world against the traditional and orthodox belief that each kind of species was created separately by God in the beginning.

Von Humboldt's book having awakened the desire to visit the tropics, William Edward's *A Voyage Up the River Amazon* suggested the destination, intensifying "an earnest desire to visit a tropical country, to behold the luxuriance of animal and vegetable life said to exist there, and to see with my own eyes all those wonders which I had so much delighted to read of in the narratives of travellers."[5] And so it happened that Alfred Wallace, accompanied by his friend, the entomologist Henry Bates, boarded HMS *Mischief,* and sailed to Brazil in 1848. Unlike Darwin, they were not the guest of the Royal Navy, but had to pay for their voyage from their own pocket. "Compared with their counterparts on naval survey ships, these two penurious, inexperienced young men faced almost unimaginable challenges."[6] Eventually Bates would continue on his way up the Amazon, while Wallace chose to branch off on the Rio Negro, a majestic river in its own right. Later Alfred's brother Herbert joined him, but did not prove strong enough to stand the ordeals of the tough and dangerous explorer's life. He caught yellow fever and died in a Brazilian harbour town, about to board a ship to his homeland.

After four years in the rain forest, Wallace returned to Great Britain

---

3. http://en.wikipedia.org/wiki/Alfred_Russel_Wallace (emphasis added).
4. Carl Zimmer: *Evolution,* p. 46.
5. Michael Shermer: op. cit., p. 58.
6. Iain McCalman: *Darwin's Armada,* p. 235.

in August 1852. As bad luck would have it, highly flammable natural lacquer called balsam of capivi in the ship's cargo caught fire. "After its long stew in the tropical sun the old ship was as dry as a tinderbox." Its crew and passenger had to abandon ship, and the naturalist watched from a life boat how the brig went down in flames, together with the greatest part of the specimens he had collected in four strenuous years, and most of his precious notes and sketches. After ten days at sea in the open boat they were rescued and taken to a home port.

"He had no qualifications, no money, no patrons, no clothes. Nor did he have any publications or specimens to show for his four years of backbreaking work. His ankles were so swollen that he could barely walk, and his thin tropical shirt failed to keep out the October wind."[7] Still he wrote six academic papers and two books, one of which was titled *Travels on the Amazon and the Rio Negro*. The specimens he had been able to send and his writings gave him some renown. He got into contact with the likes of the geologist Charles Lyell and, most importantly, Charles Darwin.

Wallace travelled in the Malay Archipelago, now Malaysia and Indonesia, from 1854 to 1862, which means that, adding the years in the Amazon, he was active as a naturalist in the field and an explorer for no less than twelve years. He was the first European to set foot on many islands, and "he collected more than 125,000 specimens in the Malay Archipelago (more than 80,000 beetles alone). More than a thousand of them represented species new to science."[8] He reported finding an average of forty-nine new species a day, "with a high of seventy-eight in one particularly good catch." Several species are named after him.

The circumstances in which he lived there were appalling. He was often weak, sick – he had contracted malaria – and starving, not to mention poor, and on occasion barely survived. "The people here have some peculiar practices," he wrote. "'Amok', as we say 'running a muck', is common here. There was one last week; a debt of a few dollars was claimed of a man who could not pay it so he murdered his creditor, and then knowing he could be found out and punished he 'run a muck', killed four people, wounded four more and died what the natives call a honourable death! A friend here seeing I had my mattress on the floor of a bamboo house which is open beneath, told me it was very

---

7. Iain McCalman: op. cit., p. 249.
8. Wikipedia: op. cit.

dangerous as there were many bad people about who might come at night and push their spears up through me from below."[9]

The days in March 1858 which Wallace passed on the island of Ternate are written in history.

A Theory of Evolution

Darwin was knowledgeable of several ideas on 'transmutationism' or 'transformism' (the opinion that species change over time), but had left on his voyage around the world without taking any of them seriously; he was still a convinced creationist at the time, and very impressed by the arguments of natural theology, more specifically those of William Paley. Wallace had severed his ties with creationism earlier in life and was already in the tropical forests of the Amazon searching for facts that would support one evolutionary theory or the other. He was also thoroughly familiar with Lyell's writings on geology and Malthus' sensational *Essay on the Principle of Population*.

About Chambers' *Vestiges of the Natural History of Creation*, which, as we just saw, played an important part in his thinking, he wrote to Henry Bates: "I do not consider it a hasty generalization, but rather an ingenious hypothesis strongly supported by some striking facts and analogies, but which remains to be proven by more facts and the additional light which more research may throw upon the problem. It furnishes a subject for every student of nature to attend to; every fact he observes will make either for or against it, and it thus serves both as an incitement to the collection of facts, and an object to which they can be applied when collected."[10] Each one of the countless facts Wallace observed during his explorations was examined with care and tentatively placed within a framework which would explain all of them together: a theory of transmutation or transformation. (The word 'evolution' would come in vogue around 1870.)

Alfred Wallace was the founder of 'biogeography', the distribution of plants and animals in the natural environments of the Earth. It had become clear to him that species had their specific habitat, generally separated by natural barriers – seas, mountains, large rivers. Islands demonstrated this fact quite convincingly, even when close to each

---

9. Michael Shermer: op. cit., p. 109.
10. Wikipedia: op. cit.

other. In the Galapagos, Darwin had made a similar observation, which would eventually lead to his theory. Wallace found out that in the Malay Archipelago he could draw a line separating a group of western islands from the eastern ones: on the western side the animals were related to the Asian species, on the eastern to the Australian. (This line, slightly modified, is still called "the Wallace Line.")

From 1852 to 1855, while in the field, Wallace's reflections on a comprehending evolutionary framework became more coherent. We know that the idea of 'transmutation' of the species was not new to naturalists, and Wallace was doubtlessly acquainted with most of the proposed explanations. Yet the problem, "the mystery of mysteries," was the mechanism: *how* did the change in the species, the formation of new species happen? In 1855, Wallace wrote an article which was a first approach to a solution: "On the Law which has Regulated the Introduction of New Species." The article was published in a scientific periodical, *Annals and Magazine of Natural History*, and is now famous as "the Sarawak paper," named after the island in which it was written. Its main statement was that "every species has come into existence coincident both in space and time with a pre-existing closely allied species."[11] Its direct inspiration was Lyell's geological theory of gradualism and Wallace's personal work in biogeography. The meaning at the core was no less than gradual evolution. Wallace declared that "closely allied species came into existence not only near one another in space, but *from* one another in *time*."

The esteemed Charles Lyell read the article and was so impressed by it that he sent a warning to his friend Charles Darwin: somebody else seemed to be closing in on a theory of natural selection. It should be noted that at that time, after having pondered the subject for no less than twenty years, Darwin still kept his theory a secret even from his closest friends. Now he wrote to Wallace: "I can plainly see that we have thought much alike and to a certain extent have come to similar conclusions." But he deemed himself quite superior to the practically unknown man in Sarawak, and wrote in the margin of his copy of the article "nothing very new." – "Alfred Wallace was someone Darwin had never taken seriously, even in the face of warnings."[12]

Wallace was suffering from malaria, which struck him with severe bouts of fever. It was on March 1858, during one such attack in the

---

11. Michael Shermer: op. cit., p. 85.
12. Iain McCalman: op. cit., p. 318.

island of Ternate, that illumination hit and made him see all elements he had gathered over the years in one consistent picture. Once again, just like in Darwin's case, it was Malthus who triggered the idea. "During one of these fits, while again considering the problem of the origin of species, something led me to think of Malthus' *Essay on Population* … There suddenly flashed upon me the idea of *the survival of the fittest* – that the individuals removed by these [population] checks must be, on the whole, *inferior* to those that survived. Then, considering the *variations* continually occurring in every fresh generation of animals or plants, and the changes of climate, of food, of enemies always in progress, the whole method of specific modification [i.e. change of the species] became clear to me, and in the two hours of my fit I had thought the main points of the theory."[13] That very evening he "sketched the draft of a paper" and wrote his complete theory down in two nights.

Wallace sent "On the Tendency of Varieties to Depart Indefinitely from the Original Type" to Darwin. The essay took two months by ship to reach its destination – and shocked its recipient. "Darwin sent the manuscript to Charles Lyell with a letter saying 'he could not have made a better short abstract! Even [Wallace's] terms now stand as heads of my chapters."[14] He had confided to Lyell earlier: "I rather hate the idea of writing for priority, yet I certainly should be vexed if anyone were to publish my doctrine before me."[15] Ever the recluse, and distraught by the illness of one of his children, he entrusted the problem to his friends Lyell and Hooker. Now Darwin's theory *had* to be brought into the open; if not, he would loose the priority of its formulation, for it would be conferred on Wallace.

On 1 July 1858 extracts from Darwin's writings and Wallace's essay were read at a meeting of the Linnean Society in London, in this way ascertaining that the original idea had been Darwin's. Recently several historians of science have claimed that the manoeuvre of Darwin's influential friends was a conspiracy, in which he was involved, to make him come victorious out of the contest. Robert Wright, for instance, writes: "Charles Lyell and Joseph Hooker were part of Darwin's coalition, and manoeuvred to elevate his status at the expense of Alfred Russel Wallace … Wallace had been taken to the cleaners. His name, though given equal billing with Darwin's, was now sure to be eclipsed by it. After all, it wasn't news that some young upstart had declared

---

13. id., p. 113 (emphases in the text).
14. Wikipedia: op. cit.
15. id., p. 89, 142.

himself an evolutionist and proposed an evolutionary mechanism; it *was* news that the well-known and respected Charles Darwin had done so. ...Today Darwin is Darwin, and Wallace is an asterisk." [16]

How did far-away Wallace react to that historical meeting of the Linnean Society? "He accepted the arrangement after the fact, happy that he had been included at all. Darwin's social and scientific status was far greater than Wallace's, and it was unlikely that, without Darwin, Wallace's views on evolution would have been taken seriously. Lyell and Hooker's arrangement relegated Wallace to the position of co-discoverer, and he was not the social equal of Darwin or the other prominent British natural scientists. ... This, combined with Darwin's (as well as Hooker's and Lyell's advocacy) on his behalf, would give Wallace greater access to the highest levels of the scientific community." [17]

In fact Wallace, "the man on the outside," has always remained respectful of Darwin and grateful to him. After having received a copy of *The Origin of Species,* he said to a friend: "Mr. Darwin has created a new science and a new philosophy; and I believe that never has such a complete illustration of a new branch of human knowledge been due to the labours and researches of a single man. Never have such vast masses of widely scattered and hitherto quite unconnected facts been combined into a system and brought to bear upon the establishment of such a grand and new and simple philosophy." [18] He dedicated *The Malay Archipelago,* one of the most popular books of scientific exploration, to Darwin, and called him late in life still "my honoured friend and teacher."

Shermer writes: "Wallace did not feel the loser, because he was not. An essay written in two nights, sent to the right place at the right time, put him in the scientific inner circle and into the historical record – his name next to Darwin's – forever." [19] One might beg to differ with the words "written in two nights", for Wallace's illumination and the feverish penning of his essay were the eruptive result of at least ten years of constant observation and reflection.

---

16. Robert Wright: *The Moral Animal,* pp. 275, 304, 288.
17. Wikipedia: op. cit., p. 9.
18. Michael Shermer: op. cit., p. 134.
19. Michael Shermer: op. cit., p. 150.

Wallace Breaks with Darwin

It would have been a wonder if two theories, composed by two natural scientists isolated from each other, had been similar in every detail. They certainly were the same in outline – which is what caused Darwin's stupefaction – but on some topics the accents differed. For example, "Darwin emphasized competition between individuals of the same species to survive and reproduce, whereas Wallace emphasized environmental pressure on varieties and species forcing them to become adapted to their local environment."[20] Another point of disagreement was Darwin's theory of sexual selection, which Wallace would never accept. (If natural selection is the one and only mechanism, as is still vociferously asserted, how can it ever be replaced or supplemented by another mechanism?) And a third point, mostly kept under wraps by Darwinians, was Lamarckism. For Darwin had bit by bit integrated more Lamarckian elements in his revised editions of the *Origin*, while Wallace remained squarely opposed to the "acquired characteristics" of the Frenchman.

The break with Darwin happened in 1869, ten years after the publication of the *Origin*. According to Wallace "certain of our physical characteristics" were not explicable by the theory of variation and survival of the fittest, the backbone of Darwinism. "These [characteristics] include the human brain, the organs of speech and articulation, the human hand, and the external human form, with its upright posture and bipedal gait."[21] What Darwin could not explain, in short, was the evolution of the human mind and its physical instrumentation. "How then was an organ [the brain] developed far beyond the needs of its possessor?" Wallace asked. "Natural selection could only have endowed the savage with a brain a little superior to that of an ape [according to the gradualism of Darwinian evolution], whereas he actually possesses one but very little inferior to that of the average members of our learned societies."[22]

"As Wallace observed: raised in England instead of the Ecuadorian Amazon [well-known to him], a native child of the head-hunting Jivaro, destined otherwise for a life spent loping through the jungle, would learn to speak perfect English, and would upon graduation from

---

20. Wikipedia: op. cit.
21. David Berlinski: *The Devil's Delusion,* p. 157.
22. Michael Shermer: op. cit., p. 159.

Oxford or Cambridge have the double advantage of a modern intellectual worldview and a commercially valuable ethnic heritage ... From this it follows, Wallace argued, that characteristic human abilities must be latent in primitive man, existing somehow as an unopened gift, the entryway to a world that primitive man does not possess and would not recognize."[23] The problem of the evolution of the human brain and its capacities has recently become trendy because of the much publicized theory of "evolutionary psychology." According to this view of our species, for ninety-nine percent of human existence people have lived as foragers in small nomadic bands, which resulted in our brains being adapted to that long-vanished way of life. Consequently, the modern mind is fit for the Stone Age, not the computer age.

As an evolutionary scientist, Wallace raised another pertinent question in connection with the mind-body problem: "How are these physical processes connected with the facts of consciousness? The chasm between the two classes of phenomena would still remain intellectually impassable." Consciousness, Wallace argued, is a qualitative phenomenon, not a quantitative one. It cannot be spontaneously generated by adding more molecules, as if there were some critical mass that, when reached, produces consciousness. "If a material element, or a combination of a thousand material elements in a molecule, are alike unconscious, it is impossible for us to believe that the mere addition of one, two, or a thousand other material elements to form a more complex molecule could in any way tend to produce a self-conscious existence."[24] Since then libraries have been written about the mind-body problem, but materialistic science is still unable to solve it.

In Wallace's writings one began now to find passages mentioning a "purpose" in nature, which was anathema to scientific materialism. "There is purpose, then, in what is, and in what happens in Nature." He also mentioned an "Overruling Intelligence." But he kept repeating that his conclusions were made exclusively on the basis and in the spirit of science. Even today one hardly ever reads a page about living organisms without hitting on a teleological metaphor, but that there should be a trend in the evolution is always curtly if not sarcastically denied. Matter is matter and cannot have a purpose, and neither can a mechanism consisting of matter. In science the metaphor of the machine is a persistent one, however refined it has become in our micro-technological age; it

---

23. David Berlinski: op. cit., pp. 158-59.
24. Michael Shermer: op. cit., p. 170.

seems for the time being the only one to enable humankind to reduce, understand, and construct.

Darwin's reaction to Wallace's apostasy from scientific orthodoxy was bitter, perhaps more so because Charles Lyell, at last accepting evolution after many years of friendship with Darwin, seemed to lean over to Wallace. "These views greatly disturbed Darwin, who argued that spiritual appeals were not necessary and that sexual selection could easily explain apparently non-adaptive phenomena." (Darwin's reaction reminds of Sigmund Freud's when Carl Jung tore himself loose from him. Maybe it is typical for all founders of a faith when the beloved disciple turns his back on them.) Many other members of the scientific elite joined Darwin is his rejection of Wallace's deviation, "for his views were at odds with two major tenets of the Darwinian philosophy, which were that evolution was not teleological and that it was not anthropocentric."

Wallace would be the most cited naturalist, though often in strong disagreement, in Darwin's *Descent of Man* (1872). Wallace himself, more broad-minded, continued respecting "his master and friend" and his work. In 1889 he even published a much cited book, called *Darwinism*, in response to criticism among scientists of natural selection. In those years the fact of evolution was already widely accepted in scientific circles and among the educated public – although Wallace and August Weismann (to be met with later) "were nearly alone among the prominent biologists in believing that natural selection was the major driving force behind it."[25] For rough times awaited Darwin's Darwinism.

'Intelligent Design'

Readers informed about the present controversy concerning 'intelligent design' will have noticed that Alfred Wallace was stepping on the same path. He did so openly, sincere and outspoken as always, and made his viewpoint perfectly clear. He reminded his readers of the terms he had used for the Overruling Intelligence – "some other power", "some intelligent power", "a superior intelligence", "a controlling intelligence" – and the way he had, on the basis of observation and well-informed, rational thought, come to such a conclusion.

He had already written in 1856: "Many animals are provided with organs and appendages which serve no material or physical purpose. The

---

25. Wikipedia: op. cit.

extraordinary excrescences of many insects, the fantastic and many-coloured plumes which adorn certain birds, the excessively developed horns in some of the antelopes, the colours and infinitely modified form of many flower-petals, are all cases for an explanation of which we must look to *some general principle far more recondite* than a simple relation to the necessities of the individual." No one, layman or scientist, can admire the creatures of nature and wonder at them without occasionally having the same impression. But "naturalists are too apt to imagine, when they cannot discover, a use for everything in nature," remarked Wallace.[26] Observers today are of the same opinion, e.g. Michael Behe: "Some evolutionary biologists ... have fertile imaginations. Given a starting point, they almost always can spin a story to get to any biological structure you wish."[27]

Still, by 1867 Wallace was "an uncompromising adaptationist, a hyper-selectionist." And he will write many years later: "Although I maintain, and even enforce my differences from some of Darwin's views, my whole work tends forcibly to illustrate the overwhelming importance of Natural Selection over all other agencies in the production of new species ... Even in rejecting [Darwin's] phase of sexual selection depending on female choice, I insist on the greater efficacy of natural selection. This is pre-eminently the Darwinian doctrine, and I therefore claim for my book [*Darwinism,* 1889] the position of being the advocate of pure Darwinism."[28]

"Pure Darwinism" meant scientific materialism. Then how could Wallace write "an Overruling Intelligence has watched over the action of those laws [of nature], so directing and so determining their accumulation as finally to produce an organization sufficiently perfect to admit of, and even to aid in, the indefinite advancement of our mental and moral nature"? The conclusion seems to be that he, less prejudiced than doctrinaire scientific materialists, saw *everything* as Nature, as a form of manifested matter in action. Matter was the fact, matter was the mystery. His rational mind would not accept anything he could not explain, and held that in Nature everything had to be mechanism in one way or other. The Chain of Being – matter, life-force, mind – was no longer taken into consideration in the second half of the nineteenth century. Therefore Wallace had to consider, even when denying it, a *materia mystica* with the power to perform the wonders of nature.

---

26. Michael Shermer: op. cit., p. 209 (emphasis added).
27. Michael Behe: *Darwin's Black Box,* p. 65.
28. Michael Shermer: op. cit., p. 211.

This is shown by his enumeration of three stages in the Earth's past that cannot be accounted for by natural selection: 1. "the change from inorganic to organic, when the earliest vegetable cell, or the living protoplasm out of which it arose, first appeared"; 2. "the introduction of sensation or consciousness"; 3. "the existence in man of a number of his most characteristic and noblest faculties," namely those of the mind.[29] Earth is a planet whose basis is matter; therein the life forces manifested; then therein consciousness has gradually lit up. At these three points in the evolution on our Earth "the unseen universe of the Spirit" has interceded, averred Wallace.[30] The two axis points in between the three phases of the evolution are precisely the points of insertion where a higher level of the hierarchical Chain of Being became active in the existence of our planet.

Spiritism

At the time of Wallace's break with Darwin an important change must have taken place in his mind. And indeed it had: Wallace had accepted spiritism as genuine. "When I returned from [the Malay Archipelago] I had read a good deal about spiritualism, and, like most people, believed it to be a fraud and a delusion. This was in 1862. At that time I met Mrs. Marshall, who was a celebrated medium in London, and after attending a number of their meetings, and examining the whole question with an open mind and with all the scientific application I could bring to bear upon it, I came to the conclusion that spiritualism was genuine. However, I did not allow myself to be carried away, but I waited for three years and undertook a most rigorous examination of the whole subject, and was then convinced of the evidence and genuineness of spiritualism." (Wallace, in the Anglo-Saxon fashion, uses the word 'spiritualism' for the phenomena connected with the evocation or apparition of beings from another world; as this may cause confusion with spirituality, in the present essay the word 'spiritism' is used, except in direct quotations which have 'spiritualism'.)

Nonetheless, Wallace consistently and persistently considered himself a scientist in the unadulterated sense of the word. Shermer writes: "Wallace's belief in spiritualism was based on a rational, scientific analysis of

---

29. id., p. 173.
30. Wikipedia: op. cit.

the phenomena, not on blind faith, typically associated with religious devotion." And Wallace himself affirms that in the years before his discovery of spiritism, "I was so thorough and confirmed a materialist that I could not at that time find a place in my mind for the conception of spiritual existence, or for any other agencies in the universe than matter and force. Facts, however, are stubborn things. ... My desire for knowledge and love of truth forced me to continue the inquiry. The facts became more and more assured, more and more varied, more and more removed from anything that modern science taught, or modern philosophy speculated on. The facts beat me."[31]

In Wallace's view "there is no supernatural. There is only the natural and unexplained phenomenon yet to be incorporated into the natural. It was one of Wallace's career goals to be the scientist who brings more of the apparent supernatural into the natural." (Shermer[32]) It is interesting to see how Wallace, while trying to encompass all of Existence in his perception of Reality, rediscovers the existence of the gradations of being, which he calls "the law of continuity." "The incapacity of the modern cultivated mind to realize the existence of any higher intelligence between itself and Deity, angels and archangels, spirits and demons, have been so long banished from our belief as to have become actually unthinkable as actual existences, and nothing in modern philosophy takes their place. Yet the grand law of 'continuity', the last outcome of modern science, which seems absolute throughout the realms of matter, force, and mind, so far as we can explore them, cannot surely fail to be true beyond the narrow sphere of our vision, and leave an infinite chasm between man and the Great Mind of the universe."[33]

Wallace did not adhere to any traditional religion and he did not believe in a personal God. He considered his exploration of spiritism and the supernatural as purely and exclusively scientific, on scientific principles and in the spirit of experimental science as practiced since Galileo Galilei. But what was fancy and superstition to other scientists, Wallace found to be factually true according to his experience as a scientist, but as yet insufficiently known or understood. "A century ago," he wrote in support of his view, "a telegram from 3000 miles distance, or a photograph taken in five seconds [*sic*], would not have

---

31. Michael Shermer: op. cit., p. 164-5, 199.
32. id., p. 173.
33. Michael Shermer: op. cit., p. 172.

been believed possible, and would not have been credited on testimony, except by the ignorant and superstitious who believed in miracles."[34]

A Revival of Occultism

The nineteenth century is generally characterized as the age of dry positivism, bourgeois commercialism and science. But its second half was also the time of an astounding revival of occultism in its various aspects. Spiritism was one of the most important. The publication in 1857 of the *Livre des Esprits* (book of the spirits) by Allan Kardec, ere long called "the pope of spiritism," resulted in the emergence of a real Church with millions of followers, and of which the ministers were mostly women, the mediums. The movement spread like wildfire in the West.[35] At its core was something quite different from the hysterical sensationalism as it is so often caricatured. Spiritism was the expression of an existential need in the human being, of an activation of its various levels of being, immensely richer than the sole material level to which scientific thought reduced the world.

About this occult revival the historian Nicholas Goodrick-Clarke writes: "Occult science tended to stress man's intimate and meaningful relationship with the cosmos in terms of 'revealed' correspondences between the microcosm and macrocosm, and strove to counter materialist science, with its emphasis upon tangible and measurable phenomena and its neglect of invisible qualities respecting the spirit and emotions. These new 'metaphysical' sciences gave individuals a holistic view of themselves and the world in which they lived. This view conferred both a sense of participation in a total meaningful order and, through divination, a means of planning one's affairs in accordance with this order."[36]

Spiritism was considered no less than a new religion by its adherents. "The question of the continuation of life and the hope that death at the end is actually not the end are too deeply anchored in the human being not to try to respond to them. ... It is for the most part people thirsting for instruction and enlightenment who gather in what one might call a 'circle'. The official way of the Christian Churches and their regular

---

34. id., p. 181.
35. See e.g. Louis Pauwels and Guy Breton: *Histoires magiques de l'histoire de France,* pp. 253 ff.
36. Nicholas Goodrick-Clarke: *The Occult Roots of Nazism,* p. 29.

sermon on Sunday, which the mass of those present lets passively go over their heads without actually listening to it, does not satisfy them any longer. Their living spirit demands more nourishment than that."[37]

During the same period, and often closely connected with spiritism, there was "the rebirth of magic,"[38] with authors like Éliphas Lévi and Papus who are still being reprinted; there was an expansion of the masonic movement, which is a form of structured and hierarchized occultism; a spreading of Rosicrucianism, closely related to freemasonry; and the foundation of the rapidly proliferating theosophical movement in 1875 – not to mention alchemy, mesmerism (i.e. hypnotism), the illuminati, and dark satanic sects ... A glance at the literature of that time will meet with Honoré de Balzac (a disciple of Swedenborg), Victor Hugo, a practicing spiritist, Joris-Karl Huysmans and his novels about black magic, Edgar Allan Poe, Arthur Conan Doyle and his idiosyncratic Sherlock Holmes, Marcel Proust, chronicler of his time, and the superlative French poets Charles Baudelaire, Arthur Rimbaud, Stéphane Mallarmé and Paul Valéry. And there were of course the Impressionists, scandalizing but revolutionizing the world of the arts, not to forget the philosophers Friedrich Nietzsche, loosing his mind in the whirlpool of the age, Henri Bergson, and Sigmund Freud.

This cascade of names reminds us that the age of Jules Verne (1828-1905), contemporary of Alfred Wallace, was much more than a time of rigidity and strictness *à la* Queen Victoria and the French *messieurs* in their cylindrical top hats and *mesdames* in their crinolines. This was an explosive age, the more so for the pressure exerted by the formal restrictions on its surface. It was in fact the end of the centuries' long transition from the Middle Ages to an age yet to come, announced and prepared by the wars and global upheavals of the twentieth century – an age of which our Earth as a whole is in labour.

Spiritism and Science

In the context of our examination of the importance of Alfred Wallace, however, the most important side of spiritism was that it saw itself as *scientific*. Wallace wrote: "The speculations [of spiritism] are

---

37. Moritz Bassler and Hildegard Châtellier (ed.): *Mystique, mysticisme et modernité ...*, pp. 95, 96.
38. See the book of the same title by Francis and Isabel Sutherland.

usually held to be far beyond the bounds of science; but they appear to me to be more legitimate deductions from the facts of science than those which consist in reducing the whole universe to matter conceived and defined so as to be philosophically inconceivable." Camille Flammarion, astronomer and spiritist, wrote: "Spiritism is not a religion, it is a science of which we hardly know the ABC. Physical science teaches us that we live in the middle of a world which is invisible to us, and that it is not impossible that there are beings (equally invisible to us) who also live on the Earth in an order of perception totally different from ours."[39]

After all, this was the time of William Crookes' cathode rays, Wilhelm Röntgen's X-rays and the mysterious radiation discovered by Henri Becquerel. After the revolution introduced by the steam engine, electricity was working its wonders. People could now communicate over long distances by telegraph and telephone. And were the invisible rays, of which so many variants were being discovered, any less occult than the visitors from beyond death and the strange matter in which they appeared?

"The magic came out of the laboratories," not only out of the occult séances. In those days "the scientists were still incapable to explain how waves with mysterious capacities could cross oceans and continents, and how they could be received"[40] – as they were incapable of explaining the occult phenomena, or explaining them away. They could only accept or refuse them because they liked or disliked them. Several of the greatest names in science showed active interest in spiritism, among them Crookes and Flammarion, already mentioned, Pierre and Marie Curie (Nobel Prize winners), Charles Richet (Nobel Prize winner), the mathematician Augustus de Morgan, Cesare Lombroso, the physicist Lord Rayleigh, and others.

But what about Charles Darwin, how did he react to the spiritist wave carrying Alfred Wallace away from him? It so happened that the Recluse of Down let himself be prompted to attend a séance, but "I found it so hot and tiring that I went away before all these astounding miracles, or jugglery, took place. ... The more I thought of all that I had heard happened at Queen Anne St., the more convinced I was it was all imposture."[41] This was nothing but sheer prejudice of a mind who had

---

39. Bernadette Bensaude-Vincent and Christine Blondel (ed.): *Des savants face à l'occulte*, p. 94.
40. Bernadette Bensaude-Vincent and Christine Blondel (ed.): op. cit., p. 11, 12.
41. Michael Shermer: op. cit., p. 183.

decided beforehand that all that was nothing but "astounding miracles," "jugglery" and "imposture," without having the patience to witness it at least once. Which is neither in accordance with the scientific method, nor with Bacon's inductive method, nor with the hypothetico-deductive method.

In this, as in so many other matters, Darwin set the tone for his future disciples, who quote him not only as a naturalist but as a source of ultimate wisdom, which he certainly was not. In a case about spiritism brought to justice, Wallace testified for the defense, Darwin for the prosecution. Shermer even writes that Darwin waged "a secret war on spiritualism," hiding his weighty involvement. He said of himself: "I fear I am a wretched bigot on the subject."[42] (That Darwin was capable of this kind of attitude strengthens the suspicion that, indeed, in 1858, when the priority of the theory of evolution by natural selection was settled in his favour, Wallace may have been the victim of a conspiracy with Darwin's knowledge.) Darwin will be joined by the scientific establishment, in his time and in the eyes of the future, to condemn Alfred Wallace as scientifically spurious, and thereby eclipse him to the present day. As Joseph Hooker wrote to his friend Charles: "Wallace has lost caste terribly,"[43] and he has not yet regained it, although he amply deserves regaining it.

"Occultism is in its essence man's effort to arrive at knowledge of secret truths and potentialities of Nature which will lift him out of slavery to his physical limits of being," wrote Sri Aurobindo, the Indian philosopher and yogi. "This human aspiration takes its stand on the belief, intuition and intimation that we are not mere creatures of the mud, but souls, minds, wills that can know all the mysteries of this and every world, and become not only Nature's pupils but her adepts and masters." However, "occultism in the West could be thus easily pushed aside because it never reached its majority, never acquired ripeness and a philosophic or sound systematic foundation. It indulged too freely in the romance of the supernatural or made the mistake of concentrating its major effort on the discovery of formulas and effective modes for using supernatural powers. It deviated into magic white and black or into romantic or thaumaturgic paraphernalia of occult mysticism and the exaggeration of what was after all a limited and scanty knowledge.

---

42. id., pp. 188, 197, 198.
43. Michael Shermer: op. cit., p. 274.

These tendencies and this insecurity of mental foundation made it difficult to defend and easy to discredit, a target facile and vulnerable."[44]

But Sri Aurobindo, with his cultural roots in the West as well as in the East, writes also about the same subject: "To know of these [occult] things and to bring their truth and forces into the life of humanity is a necessary part of its evolution. Science itself is in its own way an occultism; for it brings to light the formulas which Nature has hidden and it uses its knowledge to set free operations of her energies which she has not included in her ordinary operations and to organize and place at the service of man her occult power and processes, a vast system of physical magic – for there is and can be no other magic than the utilization of secret truths of being, secret powers and processes of Nature. It may even be found that a supraphysical knowledge is necessary for the completion of physical knowledge, because the processes of physical Nature have behind them a supraphysical factor, a power and action mental, vital or spiritual which is not tangible to any outer means of knowledge."[45] Alfred Wallace seems to have had an intuition of a kindred idea.

The loss of the knowledge and practice of true spirituality in the West has gradually led to the negation of anything non-material, to gross materialism. And rightly so! The civilization of the European Middle Ages was, like *all* other civilizations, based on imagination, myth and superstition, covering a modicum of real spirituality. The long and painful process of the Renaissance, Enlightenment, industrial revolution and materialism, resulting in the explosion of the twentieth century wars and globalization, has led humanity to more reality, to a nearer approach of truth. The present, postmodernist time, when even the possibility of any knowledge of reality or truth is negated, may be the moment when humanity discovers that the Truth is other than supposed until now.

"To refuse to enquire upon any general ground preconceived and *a priori* is an obscurantism as prejudicial to the extension of knowledge as the religious obscurantism which opposed in Europe the extension of scientific discovery. The greatest inner discoveries ... cannot be brought before the tribunal of the common mentality which has no experience of these things and takes its own absence or incapacity of experience as a proof of their incapacity or their non-existence. Physical truth of

---

44. Sri Aurobindo: *The Life Divine,* p. 875-76.
45. Sri Aurobindo: op. cit., p. 652.

formulas, generalizations, discoveries founded upon physical observation can be so referred, but even there a training of capacity is needed before one can truly understand and judge; it is not every untrained mind that can follow the mathematics of relativity or other difficult scientific truths or judge the validity either of their result or their process." (Sri Aurobindo[46])

In conclusion, Pierre Lagrange and Patrizia d'Andréa see the historical importance of the relation between occultism and science as follows: "The occult, presumed to be opposed to the sciences, is in fact profoundly defined by its confrontation with science. The occult is not a form of the irrational, but an active participation in the definition of the borderline between the rational and the irrational. The people involved do not stand on one side or the other of the Great Divide [between occultism and science], but establish this divide by the social position they take up and the definitions they formulate. Far from being a pseudo-science, occultism has without any doubt made one of the most important contributions to the historical creation of a rationalist discourse on the sciences."[47] The British Society for Psychical Research was founded in 1882, the American Society for Psychical Research in 1885; a "Spiritist and Spiritualist Congress" took place at Paris in 1889.

Humanist – Socialist

If Darwin thought that bringing his theory of evolution in the open was "like confessing a murder," it was because he realized that it would be understood as a degradation of the human being from a creature made in the image of God to just an animal among animals. His disciples who are our contemporaries do not miss any occasion to remind their readers, with obvious gusto, of what is according to Darwinism the real, scientific position of the human species.

Burton Guttman, for example, writes: "*Homo sapiens* is a mammal and a primate, a member of the Class Mammalia and the order Primates that includes the monkeys, apes, and their kin. … The forces of evolution that operate on other kinds of organisms have shaped humanity just as inexorably, and they continue to do so today, however slowly."[48]

---

46. id., p. 650.
47. Pierre Lagrange and Patrizia d'Andréa: "Définitions occultes", in *Des Savants face à l'occulte,* p. 19.
48. Burton Guttman: *Evolution,* p. 141.

John Gribbin's evaluates the human as follows: "There is no reason to single out the human line as special, except for our chauvinistic interest in it. ... There is no way in which we can claim to be 'better' than *Aegyptopithecus*, or the Miocene apes, only different. They were well adapted to the world in which they lived, and we are well suited to the world in which we live."[49] One could fill a volume with similar and even more denigrating statements, like Scott Atran's: "Human beings are accidental and incidental products of the material development of the universe, almost wholly irrelevant and readily ignored in any general description of its functioning."[50]

On the contrary Alfred Wallace, the other Darwin, saw quite valid reasons to single out the human species as special. One reason, fundamental, was that a theory of natural selection could not explain the phenomena and capabilities of the human brain, which was why Wallace distanced himself from Darwin, as we have seen earlier. Wallace argued that man is not just a physical being, not just an animal, but "a duality, consisting of an organized spiritual form ... with glorious qualities which raise us so immeasurably above our fellow animals."[51] His view was "that there is a difference in kind, intellectually and morally, between man and other animals," and he dedicated a whole book to the subject: *Man's Place in the Universe*.

Wallace saw the human being as a duality: an immortal spirit in a mortal body. Death effects no change in the spirit, and it was with such spirits that a contact could be established from our material world of mortals. The reason that he refused to believe in reincarnation may have been that he understood it as metempsychosis, an aimless transmigration of the soul in bodies of animals as well as humans. The misunderstandings in connection with this topic are endless, and seeing Peter O'Toole, when in India for the filming of *Kim*, dart around on the set shouting: "I want to come back as a snake!" did not reduce the confusion. Sri Aurobindo gives this rationale for reincarnation: "All the secret of the circumstances of rebirth [i.e. reincarnation, not metempsychosis] centres around the one capital need of the soul, the need of growth [in life after life], the need of experience; that governs the line of its evolution and all the rest is accessory."[52]

This is a view which Wallace might have accepted, for he held that

49. Mary and John Gribbin: *Being Human*, p. 119.
50. John Brockman (ed.): *What is Your Dangerous Idea?* p. 173.
51. Michael Shermer: op. cit., p. 199, 231.
52. Sri Aurobindo: op. cit., p. 815.

life on Earth "is the school for the development of the spirit."[53] Behind the constantly proclaimed "progress" by which the nineteenth century was supposed to be driven superficially, he perceived an inner urge which would make humanity evolve thanks to "the inherent perfectibility of man." This was miles away from Darwin, whose brooding view of life looms over every page of his *Origin*, except for one upbeat note in the very last paragraph of the book: "There is grandeur in this view of life ..." The way this single phrase is used time and again by Darwin's hardcore disciples, when singing the praises of science in general and biology in particular, is grotesque – especially when considering the sadomasochistic pleasure with which they declare evolution to be meaningless and humans to be complete animals. (All the same, in the writings which label the human being as nothing but an animal, it is sometimes difficult to dispel the impression that the learned author means the whole of humanity, except himself.)

Bright-minded Wallace, a genius if Darwin was one, never tried to overcome his humble working-class attitude, even when much honoured later in life. Darwin remained smugly embedded in his upper-class security and let the world take care of itself, dedicating himself exclusively to his doctrine and his family. Wallace, from his early youth influenced by the progressive humanitarian socialism of Robert Owen, wanted to collaborate towards a better world. One should keep in mind the grey, miserable masses of the "proletariat" in the shanty towns of the industrial revolution, especially in Great Britain but not only there. Wallace was concerned about the Earth and humanity as a whole, this special species with qualities far superseding those of its nearest relatives in the evolutionary tree of life. He even envisioned a world inhabited by "a single homogeneous race," and, indefatigable as always, founded a utopian community – one of many at the time – called *Freeland*.

Alfred Wallace has been called "the grand synthesizer" and "the last great Victorian," among other appreciative formulas and epithets. "In courage of opinion Wallace was without peer," writes Michael Shermer. "He was ashamed of nothing and prejudiced only against those he perceived to be dogmatically close-minded to what he believed to be unambiguous factual proof of a remarkable phenomenon."[54] His death in 1913, at 90 years of age, was widely reported in the press. *The New York Times* called him "the last of the giants belonging to that wonderful

---

53. Michael Shermer: op. cit., p. 23.
54. Michael Shermer: op. cit., p. 198.

group of intellectuals that included, among others, Darwin, Huxley, Spencer, Lyell and Owen, whose daring investigations revolutionized and evolutionized the thought of the century."

"Despite this, his fame faded quickly after his death. For a long time he was treated as a relatively obscure figure in the history of science. A number of reasons have been suggested for this lack of attention, including his modesty, his willingness to champion unpopular causes without regard for his own reputation, and the discomfort of much of the scientific community with some of his unconventional ideas. Recently, he has become a less obscure figure with the publication of several biographies about him and anthologies of his writings."[55] May his exemplary presence increase.

---

55. Wikipedia: op. cit.

6.

The Chain of Being

Thrice Vishnu paced and set his step uplifted out of the primal dust;
three steps he has paced, the Guardian, the Invincible, and from
beyond he upholds their laws.

RIG VEDA

Levels of Being

When did life originate on planet Earth? Traces of one-celled organisms have been dated at 3.5 billion years, and the current estimate of the origin of life is 3.85 billion years ago, which is early as the Earth is now thought to be 4.55 million years old.

How did life appear on planet Earth? This seems to be a quite different question, even of a different order. The simple answer is that science does not know. Klaus Dose writes: "More than 30 years of experimentation on the origin of life in the fields of chemical and molecular evolution have led to a better perception of the immensity of the problem of the origin of life on Earth rather than to its solution. At present all discussions on principal theories and experiments in the field either end in stalemate or in a confession of ignorance."[1] "Scientific explanations flounder and possibilities multiply when we ask how the first cell arose on earth. Competing theories abound – which seems always the case when we know very little about a subject. Some theories, of course, come labelled as The Answer. As such they are more properly classified as mythology than as science," writes Robert Shapiro in his much-appreciated *Origins.*[2]

The origin of life is of course one of the most important problems – if not *the* most important one – in any theory of evolution. Charles

---

1. Michael Behe: *Darwin's Black Box,* p. 168.
2. Robert Shapiro: *Origins,* p. 13.

Darwin, cautiously, left it untouched. But in the process of evolution there is also that other important problem: how can the mind act upon and through the matter of the brain? According to Larry Witham this is "the mind-boggling question that divides brain theorists and philosophers alike: how can the quality of 'mind' [whatever this may be] exist at all in matter?"[3] "The enigmatic relation between conscious experience and the physical world, commonly known as the mind-body problem, has frustrated philosophers at least since Plato, and now stonewalls scientists in their attempt to construct a rigorous theory. ... Evidence is mounting that the mind-body problem is surprisingly hard and requires revision of deeply held presuppositions ... We have yet to see our first genuine scientific theory of the mind-body problem." (Donald Hoffman[4])

As to the second problem, materialistic science has cut this Gordian knot by proclaiming that, as everything is matter, the brain and all its functions, including the mind, are matter too. Nobody has yet explained, though, how matter can be conscious. René Descartes called the mind "an epiphenomenon" of the brain, which was activated by the mind through the pineal gland. He was certainly a first-rate figure in the history of philosophy and mathematics, and as he is the intellectual patron saint of France, one better thinks twice before poking fun at his pineal gland and his vortices. Since then, materialistic science has not found anything resembling a solution of the mind-body problem. This being the case, it is encouraging to read that "it is probably safe to say that by 2050 sufficient knowledge of biological phenomena will have wiped out the traditional dualistic separations of body/brain, body/mind and brain/mind."[5] Provided that in 2050 scientific materialism still rules.

First there was matter, then there was life, then there was mind. "I suppose a matter-of-fact observer, if there had been one at the time of the unrelieved reign of inanimate Matter in the earth's beginning, would have criticized any promise of the emergence of life in a world of dead earth and rock and mineral as an absurdity and a chimera; so too, afterwards he would have repeated this mistake and regarded the emergence of thought and reason in an animal world as an absurdity and a chimera," wrote Sri Aurobindo in a letter.[6]

---

3. Larry Witham: *By Design: Science of [and the Search for] God*, p. 210.
4. John Brockman (ed.): *What Are You Optimistic About?* p. 279.
5. Floyd E. Bloom: *Best of the Brain from Scientific American*, p. 66.
6. Sri Aurobindo: *Letters on Yoga I*, p. 8.

The reader may remember Wallace's crucial statement, the outcome of many years of experience and reflection: "There are three stages that cannot be accounted for by natural selection: 1. the change from inorganic to organic ... [i.e. the origin of life]; 2. the introduction of sensation or consciousness [i.e. life and the first glimmer of mind]; 3. the existence in man of a number of his most characteristic and noblest faculties such as mathematical reasoning, aesthetic appreciation, and abstract thinking [i.e. the full-blown mind of Homo sapiens]. Wallace asserted his view of the levels of being explicitly: "The grand law of 'continuity', the last outcome of modern science, which seems absolute throughout the realm of matter, [life-]force and mind, so far as we can explore them, cannot surely fail to be true beyond the narrow sphere of our vision, and leave an absolute chasm between man and the Great Mind of the universe. Such a supposition seems to me in the highest degree improbable."[7]

While Alfred Wallace, as a naturalist another Darwin but also a great human being, saw the different levels of reality as "the last outcome of modern science," scientific materialism judged this conception of his to be a fatal mistake and condemned him to oblivion for it. In official science anything that is not matter is unworthy of consideration. And everything is matter because there cannot be anything but matter – which, of course, is a circular argument. Anything else is "the survival into our day of antediluvian mysticism and superstition." (Carl Sagan) The annoying fact for the hard sciences, however, is that matter itself has become rather mysterious and is now the equivalent of energy (and vice versa), while the very material reality has escaped the physical sciences altogether since Niels Bohr, Werner Heisenberg, and their quantum mechanics.

These are fundamental problems for any person interested in his world, and in his own origin and destiny. Unfortunately, most often the polemics about them are fought out in the clouds, without much knowledge of their substance and history. This makes a short survey of the idea of the Great Chain of Being in the past useful.

---

7. Michael Shermer: *In Darwin's Shadow*, pp. 173, 172.

"The Serial Kingdoms of the Graded Law"

"Central to the perennial philosophy is the notion of the Great Chain of Being. The idea is fairly simple. Reality, according to the perennial philosophy, is not one-dimensional; it is not a flatland of uniform substance stretching monotonously before the eye. Rather, reality is composed of several different but continuous dimensions. Manifest reality, that is, consists of different grades or levels, reaching from the lowest and most dense and least conscious to the highest and most subtle and most conscious." This is Ken Wilber's definition of the chain of being in his *Eye of Spirit*. "The Absolute manifests itself in layers, dimensions, sheaths, levels, or grades – whatever term one prefers … In Vedanta these are the *koshas*, the sheaths or layers covering Brahman; in Buddhism these are the eight *vijnanas*, the eight levels of awareness, each of which is a stepped-down or more restricted version of its senior dimension; in Kabbalah these are the *sefirot*, and so on."[8]

The first part of Sri Aurobindo's major opus, *The Life Divine* – a source of Wilber's original inspiration – describes extensively the same levels of being, which are the manifestation of what he terms "Omnipresent Reality." He writes: "A solution of the whole problem of existence cannot be based on an exclusive one-sided knowledge; we must know not only what Matter is and what are its processes, but what mind and life are and what are their processes. And one must know also spirit and soul and all that is behind the material surface; only then can one have knowledge sufficiently integral for a solution of the problem."[9]

Sri Aurobindo's spiritual practice, his "spiritual realism," has always taken the sciences as known to him fully into account, and weighed the contents of their worldview against the profoundest spirituality of the past and his personal experience. "As human thought is beginning to realize, the distinction made by the intellect and the classifications and practical experiments of Science, while perfectly valid in their own field and for their own purpose, do not represent the whole or the real truth of things, whether of things in the whole or of the thing by itself which we have classified and set artificially apart, isolated for separate analysis."[10] The truth of things is not the dogmatic, deterministic domain of matter alone, but "the hieratic message of the climbing planes",

---

8. Ken Wilber: *Eye of Spirit,* pp. 39, 45 (emphases in the text).
9. Sri Aurobindo: *The Life Divine,* p. 652.
10. Sri Aurobindo: op. cit., p. 380.

"the stages of the spirit" which are the stages of consciousness, "the serial kingdoms of the graded Law," "a golden ladder carrying the Soul," "an organ scale of the Eternal's acts," "predestined stadia of the evolving Way"... – all expressions which we find in his epic poem *Savitri*, his ultimate statement.

He translated the *rik* about the original idea of the triple world in which we live, the three paces of Vishnu, from the Sanskrit of the Rig Veda as follows : "Thrice Vishnu paced and set his step uplifted out of the primal dust; three steps he has paced, the Guardian, the Invincible, and from beyond he upholds their laws. Scan the workings of Vishnu and see from when he has manifested their laws."[11] "It is difficult to suppose," he wrote, "that Mind, Life and Matter will be found to be anything else than one Energy triply formulated, the triple world of the Vedic seers." And he affirmed: "The old Vedic knowledge will be justified,"[12] not because it is old, but because it is the "perennial philosophy" formulating the truth of things – then, now, and forever – at the base of an endlessly evolving world.

The Ways of the West

The three grades or levels of being are deeply ingrained in the thought of the West. The very first time a graded world is mentioned seems to have been by Homer, who's golden chain reaches from God's throne down to the meanest worm. Plato, drawing his inspiration from Pythagoras, divided the human consciousness into three levels. The lowest was the desire soul, corresponding to the material and vegetative vital plane centered around the navel and below; higher up, purely vital and partially psychic, was the sensitive soul, centered in the heart; and the highest was the rational or actually immortal soul, centered in the heart and the head. Aristotle had a similar gradation.

It is from the ancient Greek philosophers onwards, and because of their all-permeating influence in medieval Western thought, that the real soul, the 'psyche' has been confused with the mind. The reason of this confusion seems to have been that the Greeks, despite their Mysteries, lacked the spiritual experience of the East. Descartes' basic and very influential worldview was essentially that of his Catholic educators,

---

11. id., p. 198 [Rig Veda I.22.17-21].
12. Sri Aurobindo: *Essays in Philosophy and Yoga*, p. 172.

the Jesuits. Their official teaching was a dualism of body and soul, the burdensome body being the container of the soul in a world corrupted because of Adam's Original Sin, and the immortal soul destined for heaven or hell, depending on its behaviour while in the body.

Still, behind and in tandem with the Christian duality, the scale's gradations, chiefly divided in the triad matter-life-mind, continued to determine the thinking. In science there were the material, vegetable and animal realms, the foundation of all natural classifications. Even materialistic Cartesians like La Mettrie wrote about a continuation in physical nature "from the human, down to the higher animals, down to the animals preceding these" (which shows that a descending evolution, from the most complex to the most elementary, was still a common belief among the intellectuals of the eighteenth century.) One could illustrate abundantly how the triad of the levels of being continued to shape the thinking even of a rationalist like Bertrand Russell (instinct, mind, spirit) or an out-and-out materialist like Jacques Monod (matter, emotions, thought).

Arthur Lovejoy

The Great Chain of Being (1936), the classic essay by the historian of ideas Arthur Lovejoy, is one of the books that do not age and continue influencing a culture even though being little known. "The phrase 'the Great Chain of Being' was long one of the most famous in the vocabulary of Occidental philosophy, science, and reflective poetry," writes Lovejoy, "and the conception which in modern times came to be expressed by this or similar phrases has been one of the half-dozen most potent and persistent presuppositions in Western thought. It was, in fact, until not much more than a century ago, probably the most widely familiar conception of the general scheme of things, of the constitutive pattern of the universe; and as such it necessarily predetermined current ideas on many other matters." [13]

Lovejoy posits two principles as supports of his assertion, both originating with Plato and Aristotle, who, after all, had been Plato's disciple for twenty years. The first is the principle of plenitude. The Greek reason could not imagine a gap, empty space or break in the cosmos. For the cosmos was the creation of supernal powers, for whom

---

13. Arthur Lovejoy: *The Great Chain of Being*, p. vii.

"no genuine potentiality of being can remain unfulfilled." – "Aristotle maintained that the bodies making up the cosmos were all contiguous with each other, thus composing a plenum [fullness]."[14] As Lovejoy puts it: "The perfection of the Absolute Being must be an intrinsic attribute, a property inherent in the Idea of it; and since the being and attributes of all other things are derivative of this perfection because they are logically implicit in it, there is no room for any contingency anywhere in the universe."[15] This explains the famous scholastic dictum "nature abhors a vacuum," for ceasing to be full, the world would cease to be coherent.

The second principle is the principle of continuity. "From the Platonic principle of plenitude the principle of continuity could be directly deduced. If there is between two given natural species a theoretically possible intermediate type, that type must be realized, otherwise there would be gaps in the universe."[16] From the principle of continuity, still so called by Alfred Wallace, followed logically the notion of infinitesimal gradation "which was of the essence of the cosmological Chain of Being."

"It was Aristotle," during the Middle Ages held in such high esteem that he was referred to as The Philosopher (*ille philosophus*), "who chiefly suggested to naturalists and philosophers of later times the idea of arranging all animals in a single graded *scala naturae* according to their degree of perfection. ... The result was the conception of the plan and structure of the world which, through the Middle Ages and down to the late eighteenth century, many philosophers, most men in science, and indeed most educated men were to accept without question – the conception of the universe as a Great Chain of Being."

> Vast chain of being! which from God began,
> Natures ethereal, human, angel, man,
> Beast, bird, fish, insect, what no eyes can see,
> No glass can reach; from Infinite to thee,
> From thee to nothing ...

Thus wrote the popular poet Alexander Pope (1688-1744) in his *Essay on Man*. (The wonders revealed by the microscope – the "glass" – were thrilling the intelligentsia of those days.)

Wilber writes that evolution is fully compatible with the chain of

---

14. A.C. Crombie: *Medieval and Early Modern Science,* vol. II, p. 37.
15. Arthur Lovejoy: op. cit., p. 54.
16. id., p. 58.

being, which indeed it is, and which is the reason why the classification by Carolus Linnaeus is inspired by the chain of being. Seen in this way, the chain of being lies at the basis of modern taxonomy – and it continues to structure the human perceptions in daily life and even, as mentioned in passing, the thinking of some of the most materialistic and deterministic philosophers and scientists. This too may indicate that, as long as science limits itself to the level of 'matter', reality will remain inaccessible and the fundamental problems of science – among them the origin of life and the mind-body problem – unsolved.

"Man in the Middle"

"The human being is part of the immense chain of living beings," in the view of Marcel Schutzenberger.[17] This might seem a truism if one does not recall that the classification of the human with the animals has caused a revolution in the Western conceptions, and if one does not know that Schutzenberger is one of the great living mathematicians, which renders his mentioning of the chain of being rather exceptional. But not only is man part of the chain, he seems to be positioned somewhere in the middle on it, halfway up or halfway down, higher than the animals, lower than the angels. To quote Alexander Pope again:

> Plac'd in this isthmus of a middle state,
> A being darkly wise and rudely great,
> With too much knowledge for the sceptic side,
> With too much weakness for the stoic pride,
> He hangs between; in doubt to act or rest;
> In doubt to deem himself a god or beast;
> In doubt his Mind or Body to prefer;
> Born but to die, and reas'ning but to err ...
> Created half to rise, and half to fall,
> Great lord of all things, yet a prey to all;
> Sole judge of Truth, in endless error hurl'd;
> The glory, jest and riddle of the world.

If, as Immanuel Kant also thought, "human nature occupies as it were the middle rung of the Scale of Being," the scale must consist

---

17. Jean Staune and Eric de Romain (eds.): *L'Homme face à la science,* p. 169.

not only of matter, life-force and mind, but also of spiritual gradations above the mind (which was another reason why it was and remains discarded by scientific materialism). In sum, it encompasses all Reality, "from nothing to the deity." The chain of being, writes Lovejoy, must be interpreted "so as to admit of progress in general, and of a progress of the individual not counterbalanced by deterioration elsewhere [in nature]. ... Since the scale was still assumed to be minutely graduated, since nature makes no leaps, the future life must be conceived to be – at least for those who use their freedom rightly – a gradual ascent, stage after stage, through all the levels above that reached by man here; and since the number of these levels between man and the one Perfect Being must be infinite, that ascent can have no final term. The conception of the destiny of man as an unending progress thus emerges as a consequence of reflection upon the principles of plenitude and continuity."[18]

This open-ended view, in academic literature seldom associated with former European thought, was worded by the philosopher Viscount Bolingbroke as follows: "Shall we not be persuaded rather that as there is a gradation of sense and intelligence here from animal beings imperceptible to us for their minuteness, without the aid of microscopes and even with them, up to man, in whom, though this be the highest stage, they remain very imperfect; so there is a gradation from man, through various forms of sense, intelligence, and reason, up to beings who cannot be known to us, because of their distance from us, and whose rank in the intellectual system is above even our conceptions? This system, as well as the corporeal, must have been present to the Divine mind before he made them to exist."[19]

The supposition in the last sentence refers to another major problem: if there is a chain, or ladder, or scale of being going up, how did it come about and on what is it based? The womb of all existence is the Divine, who by his creative omnipotence established the order of the world. If life followed the gradations of an ascending ladder, it could only follow the pre-established rungs of a descending one. This conception squares with the Platonic and neo-Platonic view of the 'ideas' or 'forms', which are the moulds in the consciousness of the Divine of all things existing.

Such a conception, however, might be understood as a model of a static universe, in which everything is forever established and no kind of organism, no species ever changes. It was indeed seen in this manner

18. Jean Staune and Eric de Romain (eds.): op. cit., p. 246.
19. Jean Staune and Eric de Romain (eds.): op. cit., p. 192.

in Christian theology for many centuries. But "one of the principal happenings in eighteenth-century thought was the temporalizing of the Chain of Being," writes Lovejoy. In other words, the climbing of the ladder was now conceived as a process of development in time, materialized in the gradually increasing complexity of the organisms on Earth. The intriguing fact is that this process was not supposed to stop with the human species as it is at present.

Considering all this, and with the hindsight we have, it comes no longer as a surprise that at that time the evolutionary view appeared as a logical consequence of the way of the world. It should once more be recalled that in those days the evolutionary view, called transformism or transmutationism, was dangerous and even blasphemous. Linnaeus was one of the first to have an intuition of the transmutation of the species. Change in and of the species had practically become obvious from his lifelong study of the classification of the living beings, but, like Buffon, he preferred to keep mum about it. Many of the poets and philosophers, as illustrated by a few passages quoted previously, wrote about it openly. One of the most read was Dennis Diderot, not only with his own books but with the illustrious *Encyclopédie* as his mouthpiece. During half a century there was the dance of hesitation between religiously dogmatic creationism and the scientific thesis of evolution; both standpoints, and several in between, were defended by estimable persons like Leibniz, Robinet, Erasmus Darwin and Cuvier. Till Lamarck published his theory of transformism (1800) and Darwin his theory of transmutationism (1859).

Man was no longer the lord of creation. He was now occupying the middle rung of the Scale of Being, "midway from nothing to the deity." – "The definition of him as 'the middle link' especially emphasized the peculiar duality of his constitution and the tragi-comic inner discord in him which results from this. ... The place assigned to man in the graded scale which constitutes the universe lent to this conception still greater sharpness and an air of metaphysical necessity. Somewhere in that scale there must exist a creature in which the merely animal series terminates and the 'intellectual' series has its dim and rudimentary beginning; and man is that creature." (Lovejoy[20])

In Sri Aurobindo we read: "Man is the now apparent culmination but not the real ultimate summit; for he is himself a transient being

---

20. Jean Staune and Eric de Romain (eds.): op. cit., p. 198.

and stands at the turning-point of the whole movement."[21] And: "A many-sided ignorance striving to become an all-embracing Knowledge is the definition of the consciousness of man the mental being ... Truth is relative to us because our knowledge is surrounded by ignorance ... On the surface we are still an ego figuring self, an ignorance turning into knowledge, a will labouring towards true force, a desire seeking for the delight of existence. To become ourselves by exceeding ourselves ... is the difficult and dangerous necessity, the cross surmounted by an invisible crown which is imposed on us, the riddle of the true nature of his being proposed to man by the dark Sphinx of the Inconscience below and from within and above by the luminous veiled Sphinx of the infinite Consciousness and eternal Wisdom confronting him as an inscrutable divine Maya."[22]

And he writes in *The Life Divine*: "The animal is a living laboratory in which Nature has, it is said, worked out man. Man himself may well be a thinking and living laboratory in whom and with whose conscious co-operation she wills to work out the superman, the god. ... The animal is man in the making; man himself is that animal and yet the something more of self-consciousness and dynamic power of consciousness that make him man; and yet again he is something more which is contained and repressed in his being as the potentiality of the divine: he is a god in the making. In each of these, plant, animal, man, god, the Eternal is there containing and repressing himself as it were in order to make a certain statement of his being."[23]

The Life Divine is based on the experiences of an advanced yogi and mystic, who had worked his way back to the root-knowledge of the Vedic seers. However, that this profoundest intuition and aspiration is to be found not only in the East can be exemplified, for instance, by the following lines from Edward Young (1683-1765):

> [The Sovereign Spirit of the world,]
> From the mute shell-fish gasping on the shore,
> To men, to angels, to celestial minds,
> Forever leads the generations on
> To higher scenes of being ...
> ... In their stations all may persevere
> To climb the ascent of being, and approach

---

21. Sri Aurobindo: *The Life Divine*, p. 75.
22. id., pp. 70, 565, 685.
23. Sri Aurobindo: op. cit., pp. 4, 382.

For ever nearer to the life divine.

An Idea Lives On

The economist E.F. Schumacher became world-famous because of his book *Small is Beautiful*, which propagated the important role small-scale enterprises might play in the world economy. After his death, in 1977, another essay of his was published: *A Guide for the Perplexed*. It borrowed its title from a seminal medieval treatise by the Jewish philosopher Maimonides. *A Guide* expounds the reflections of a thinker who, as in his other work, does not hesitate to disclose his original perceptions of the world.

"Our task is to look at the world and see it whole," he writes. "We see what our ancestors have always seen: a great 'Chain of Being' which seems to divide naturally in four sections – four 'kingdoms' as they used to be called – mineral, plant, animal and human. … The Chain of Being can be seen as extending downwards from the highest to the lowest, or it can be seen upwards from the lowest to the highest. The ancient view begins with the Divine and sees the downward Chain of Being as an increasing distance from the centre and a progressive loss of qualities. The modern view, largely influenced by the theory of evolution, tends to start from inanimate matter and consider man the last link of the chain, having evolved the widest range of useful qualities."[24]

To gather his material for *Uncommon Wisdom*, Fritjof Capra also interviewed Schumacher. He reports how Schumacher explained the evolutionary stratification. "Schumacher expressed his belief in a fundamental hierarchical order consisting of four characteristic elements – mineral, plant, animal, and human – with four others – matter, life, consciousness, and self-awareness – which are manifest in such a way that each level possesses not only its own characteristic element but also those of all other levels. This, of course, was the ancient idea of the Great Chain of Being, which Schumacher presented in modern language and with considerable subtlety. However, he maintained that the four elements are irreducible mysteries that cannot be explained, and that the differences between them represent fundamental jumps in the vertical dimension, 'ontological discontinuities' as he put it. 'This is why physics cannot have any philosophical impact', he repeated. 'It

---

24. E.F. Schumacher: *A Guide for the Perplexed,* p. 24.

cannot deal with the whole; it deals only with the lowest level', the level of matter."[25]

If this is true, and it is a view supported by much philosophical, scientific and spiritual evidence, the pillars of scientific materialism rest on very insecure ground. Science is the expression of the search for Truth innate in humanity. Scientific materialism, limiting itself to the material level of existence, is a dogmatic theory which will have to be revised, perhaps very shortly. This does not efface the immense mass of research and discovery accomplished through centuries by sincerely and totally dedicated persons, many of whom have sacrificed their lives for it. It has greatly profited technology, the sister of theoretical science, to shape our world. And the limitation to the material level, to bodies that can be counted, measured and weighed, was inevitable because the human intellect cannot function otherwise. It cannot grasp the reality of the whole, it can only define and work with parts of whole.

Yet, reduction of the whole to its parts, when converted into dogmatism, must ultimately lead to distortion, confusion and stunting ignorance. (The unsolved and presently unsolvable problems of the origin of life and the mind/body relation should make this clear.) Ken Wilber, widely read, has many pages in his books trying to show that the material "flatland" of science is far from the whole of reality. Two major sources of his synthetic effort are Sri Aurobindo (1872-1950) and Arthur Koestler (1905-1983). Sri Aurobindo, in The Life Divine, has provided Wilber with an extensive description of the levels of being and their interaction; Koestler has formulated the useful concept of the "holon."

"A 'part', as we usually use the word," writes Koestler, "means something fragmentary and incomplete which by itself would have no legitimate existence. On the other hand, a 'whole' is considered as something complete in itself which needs no further explanation. But 'wholes' and 'parts' in this absolute sense just do not exist anywhere, either in the domain of living organisms or of social organizations. What we find are intermediary structures on a series of levels in an ascending order of complexity: sub-wholes which display, according to the way you look at them, some of the characteristics commonly attributed to wholes and some of the characteristics commonly attributed to parts." Such entities – and anything concrete on any level of being is such an entity

---

25. Fritjof Capra: *Uncommon Wisdom*, p. 228.

– Koestler calls "holons." Holons "behave partly as wholes or wholly as parts, according to the way you look at them."[26]

Sri Aurobindo had already written: "As human thought is beginning to realize, the distinction made by the intellect and the classifications and practical experiments of Science, while perfectly valid in their own field and for their own purpose, do not represent the whole or the real truth of things, whether of things in the whole or of the thing by itself which we have classified and set artificially apart, isolated for separate analysis."[27] With these words Sri Aurobindo defines exactly the same epistemological problem Koestler and Wilber have brought to the attention of their readers, and which is crucial for all disciplines of science. At stake is the definition of an entity, any entity, which by the fact of its existence belongs to a hierarchy, i.e. at the same time to a higher and a lower order, while still being itself – the individual and the mass, the particle and the atom, the cell and the blade of grass, the animal and its species.

As Sri Aurobindo points out, the isolation of the physical unit from the totality of its existence on all levels simultaneously, is "perfectly valid" as an approach or operation of science. The reason is, as already mentioned, that otherwise the human mind would not be able to practice science, because cutting entities out of 'reality' is the way it functions. But, again, science has made this delimiting act of the mind to 'things', observable by the physical senses, into an insuperable mental dogma. As such it restricts the field of its activities to what the general public still supposes to be the material world (but what has since about a century ago become an occult world of invisible forces and even invisible matter, and recently of an infinite number of universes). This doctrine will one day have to be overcome if the search for Truth is to continue.

Arthur Koestler was one of the great intellectuals of the twentieth century, a conscious witness of his times. A witness of his times on a more restricted level, that of science, is nowadays the physicist Paul Davies. The number of popularizing books he has written and the awards he has received for this work are remarkable, taking into account that as a professor of physics he has the keep up a certain standard. The following quotes from one of his books may therefore be worthy of consideration. "There is a growing appreciation among scientists of

26. Arthur Koestler: *The Ghost in the Machine,* p. 48.
27. Sri Aurobindo: op. cit., p. 380.

the importance of structural hierarchy in nature: that holistic concepts like life, organization and mind are indeed meaningful, and that they cannot be explained away as 'nothing but' atoms or quarks or unified forces, or whatever. ... Life is a holistic concept, the reductionist perspective revealing only inanimate atoms within us. Similarly mind is a holistic concept, at the next level of description. We can no more understand mind by reference to brain cells than we can understand cells by reference to their atomic constituents."[28]

---

28. Paul Davies: *God and the New Physics*, pp. 225, 92.

7.

Inventing 'Darwinism'

Nothing in biology makes sense except in the light of evolution.

THEODOSIUS DOBZHANSKY

Darwinism on the Ropes

It is often thought that it did not take long before Darwinism was generally accepted, and that it has been sailing before the wind ever since. What *was* quite soon accepted, though in the teeth of rugged resistance, was the fact of evolution, not Darwin's theory as such. André Pichot writes in one of his thoroughly researched books: "In the years which separate us from the publication of Darwin's *Origin of Species*, there has practically not been a single moment when Darwinism was not in crisis. ... It is doubtlessly in the years around 1900 that the confusion [about Darwinism] reached its peak."[1] In his *Evolution* Edward Larson confirms this: "By the end of the nineteenth century Darwinism was on the ropes."[2]

We remember that a theory of evolution – any such theory – evoked quasi-instinctive resistances all around, in religious, non-religious and anti-religious people, and not the least in scientists, who most often had thought out their own explanations and systems, or held on to the creationist *status quo*. Lyell accepted the theory of his friend Darwin only in 1869, after having defended him for a decade. Huxley, the champion of the theory, had his reservations up to the very publication of the *Origin*, and never accepted Darwin's gradualism. Asa Gray agreed with Darwin only within a framework which held that all species had been created over time. Herbert Spencer, propagandist of Darwinism, remained a Lamarckist at heart. And of Ernst Haeckel,

---

1. André Pichot: *Histoire de la notion de gène,* pp. 275, 78.
2. Edward Larson: *Evolution,* p. 119.

"the German Darwin" and "universal promoter of Darwinism," Larson writes: "Haeckel saw evolution proceeding through the accumulation of Lamarckian acquired characteristics selected for fitness in a Darwinian fashion."[3] All these were persons close to Charles Darwin physically or in their way of thinking.

There were enough rival theories of Darwinism to keep the "X Club" occupied. The creationist theory of evolution budded into several variants: literalist, short-term and long-term, with some more theories composed according to the idiosyncratic interpretation of the Bible by their inspired authors. Then there was orthogenesis, which held that evolution was the work of an inner driving force. And Lamarckism and neo-Lamarckism were intertwined or confused with natural selection in many minds. This is understandable if one considers that the environment plays a dominant role in both theories. Each time the environment has come into play, Lamarckism has raised its head, favoured by the fact that the "mechanism" of Darwinian evolution remained problematic.

Among the major scientific problems bothering Darwin during his lifetime was the age of the Earth. William Thomson, later known as Lord Kelvin, "during most of his life widely thought of as the leading physicist and electrical engineer in the world," still thought that the Sun's energy was generated by coal. Basing himself on this supposition, he calculated an age for the Earth much too short for the evolution of life to be possible. As Darwin mentions in his *Origin*: "Sir W. Thomson concludes that the consolidation of the crust [of the Earth] can hardly have occurred less than 20 or more than 400 million years ago, but probably not less than 98 or more than 200 million years."[4] The age of the Earth as accepted at present is 4.57 *billion* years.

Another problem, still alive today, was that the mechanism of inheritance remained unknown. And a third problem was the apparent persistence of gaps or discontinuities in the fossil record. It was this last fact that gave rise to saltation theories. Now associated with the names of Stephen Jay Gould and Niles Eldredge, saltation theories opposed dogmatic gradualism already in Darwin's days. They hold that evolution has not happened gradually, tiny step by tiny step, but as it were in sudden eruptions of life forms, of which the most dramatic, as known now, was the "Cambrian Explosion" half a billion years ago.

---

3. Edward Larson: op. cit., p. 109.
4. Charles Darwin: *The Origin of Species,* p. 309.

"By the turn of the twentieth century no consensus existed among biologists about how evolution operated," writes Edward Larson.[5] "Had evolution and Darwinism indeed triumphed? Hardly. In fact, 'from about 1890 to 1910 Darwin's theory was threatened to such an extent by various opposing theories that it was in danger of going under.'"[6] The source quoted here is none other than Ernst Mayr, "dean of evolutionism" and "one of the architects of contemporary Darwinism." "In 1903 the German botanist Eberhard Dennert proclaimed: 'We are now standing by the death-bed of Darwinism, and making ready to send the friends of the patient a little money to insure a decent burial of the remains,'" concludes Edward Larson.[7] And Pichot asserts that around the turn of the century every biologist had more or less assembled his own theory of evolution.

A Hollow Theory?

"Darwinism was built bit by bit between 1859 and 1910, integrating into the original Darwinian theory various elements which were not part of it, and eliminating at least as many. The whole process came about with many difficulties and controversies, for the theses which were to be adjusted were sometimes antagonistic. ... This constant crisis was in a way normal, for Darwinism has never been able to provide evolution with a theoretical necessity, and it has always depended on external supports, with very few arguments deduced from its own premises." (Pichot[8])

As pointed out before, what was accepted was not Darwinism properly speaking, but evolution under the name of Darwinism. This acceptance was facilitated by the obvious fact that the human physiology resembles the physiology of primates quite closely, and that nature apparently manifests a hierarchy of gradations of consciousness in its mineral, plant and animal kingdoms, with other discernible degrees in the specification of living beings. There was a kind of instinctive memory of evolution in the many myths of life emerging from "the waters," e.g. in India the succession of avatars: fish, tortoise, boar, dwarf [hominid], Rama with the ax [*Homo habilis*], Rama with the bow [*Homo*

5. Edward Larson: op. cit., p. 222.
6. Hal Hellman: *Great Feuds in Science,* p. 92.
7. Edward Larson: op. cit., p. 128.
8. André Pichot: op. cit., pp. 275, 277.

sapiens], Krishna, the Buddha. Now the fossils discovered by the new sciences of geology and paleontology presented the theorists with a gradually increasing complexity which suggested transformation, and of which the relationships could be "descent with modification."

It was also the time that the myths of the Bible, which had become an integral part of the mental implementation of the Western world, could at last be openly put into question, albeit not yet without risk. The medieval, Catholic paradigm could be challenged by the paradigm of "man the rational being." Living on a fault line between two paradigms is dangerous and demands heroism, if not their lives, of the individuals who contribute to the transition consciously. The enthusiasm of such individuals, animated by the spirit of a new time, has always been admirable – e.g. the sophists in Greece, the Christian Church Fathers and martyrs, the Renaissance men of the New Learning, the philosophers of the Enlightenment.

To switch from one God, who is supposed to be non-existent or to have died, to another is a world-rending event. The "X Club", like Darwin's "four musketeers," although limited in the immediate awareness of their aims, was one of the instruments necessary to give shape to a new time, in this case the era of modern science with its own myth of scientific materialism and Darwinism. They contributed to the advent of the end of an era and the opening of the gates of a new one, not yet envisioned and therefore not yet named.

As we have seen in the chapters on Darwin, the two main pillars of his "transmutationism" were the geological theory of Charles Lyell and the economic theory of Thomas Malthus. On the unexplained fact of evolution, Darwin projected his combination of Lyell and Malthus. Geology and economics had to explain biology. According to Ernst Mayr "Darwin advanced numerous theories, among which five are most important": the non-constancy of species, which is the basic theory of evolution; the descent of all organisms from common ancestors; the gradualness of evolution (no saltations, no discontinuities); the multiplication of species; natural selection.[9]

The non-constancy of species and the descent of all organisms from other organisms are common to all naturalistic theories of evolution, also of Lamarck's transformism. Specific to Darwin are the small variations, the problem of adaptation to the environment which leads to natural selection, and the inheritance of the variations which are best

---

9. Ernst Mayr: *What Evolution is,* p. 94.

suited to the environment and therefore belong to the "fittest." How the small variations in each individual came about, Darwin had no idea, but they formed the basis on which the plant and animal breeders made their selections. How the variations were inherited, Darwin did not know, but he proposed his theory of "pangenesis" and the "gemmules," which was pure guesswork. That the fittest animals were the best to survive and breed in their environment was an experiential fact which many have called a tautology: the fittest are most fit. Darwin – and he was well aware of the fact – had put together a superficial explanation without an inner justification or substance, and therefore in point of fact hollow.

In other words, Darwin had solved the many questions cropping up in his explorations during the *Beagle* voyage, and his constant studies and interrogations of breeders and naturalist colleagues, by outlining a scheme consisting of a mixture of geology and economics, without explaining the actual "mechanism" of the evolutionary process. To use a simile: a triangle or a circle can be drawn through any three points that are not in a straight line. Taking the three essential tenets of Darwin's theory – variations, inheritance and natural selection – as three such points, Darwinians have always connected them by a circle, showing that their idol's theory is the ideal one, or by a triangle, imposing the dogmatic knowledge of Darwinism as the only valid explication of the evolution of life. But Darwinism is no more than an intellectual idealization of a fact that is mentally definable in many other ways, all of which lack the cardinal understanding of the actual mechanism – the three points can be connected by an infinite number of figures, from the simplest to the most bizarre, none of which actually *explains* the secret of the evolution of life.

In Darwinism "on the one hand, one admits that the interaction of the laws of physics does not suffice to explain a living being, and one ads to it a 'command structure' [the Darwinian dogma] which corrects this insufficiency. On the other hand, one pretends that this 'command structure' does not refer to anything that is not strictly physical, and that therefore the living being results from the interaction of the physical laws," notes Pichot. And he adds that one has to admit that the "improbable" structure of the living being, its gestation and the way it stays alive have not yet received, even today, a satisfactory physical explanation.[10]

---

10. André Pichot: op. cit., p. 28.

It is from an analysis like this one that Pichot concludes: "It is in the society of that time, and not in the biology, that one has to look for the cause of the success of Darwinism, either biological as well as social."[11] "Darwin's whole theory of evolution by natural selection bears an uncanny resemblance to the political economic theory of early capitalism as developed by the Scottish economists," writes Richard Lewontin, the best known exponent of that school being Adam Smith (1723-90). And he adds: "What Darwin did was take early-nineteenth-century *political* economy and expand it to include all of *natural* economy."[12]

Larson sees things in an analogous way: "Essential to Darwin's conception was a modern worldview influenced by ideas of utilitarianism, individualism, imperialism, and *laissez-faire* capitalism. Of course Malthus was a utilitarian-minded political economist who championed the *laissez-faire* ideal. Darwin also read the writings of Adam Smith and other utilitarian economists who presented individual competition as the driving force of economic progress. Perhaps more important, he lived in a society that embraced this view; Darwin himself came from a family of successful capitalists. Further, he rode on the rising tide of British economic, political, and cultural imperialism ... Natural selection intuitively seemed the right answer to a man thoroughly immersed in the productive, competitive world of early Victorian England."[13]

A Passion for Peas

The following sections of this chapter are more than an enumeration of names and data: they narrate the steps by which the biological sciences tried to provide experimental support for Darwin's hypothetical outline. It is the reason why this chapter is called "Inventing 'Darwinism'" – 'Darwinism' in inverted commas because, once again, the gradually assembled scientific argumentation of evolution differs considerably from the sketch of a theory of evolution as worked out by Charles Darwin.

Johann Gregor Mendel (1822-84), an Augustinian monk, was a younger contemporary of Charles Darwin. 'Johann' was the name given him at his baptism, 'Gregor' the name he chose when becoming a monk

---

11. André Pichot: *La société pure,* p. 77.
12. Richard Lewontin: *The Doctrine of DNA,* p. 10.
13. Edward Larson: op. cit., p. 70.

at the abbey of Brno, now in the Czech Republic. Towards the end of his life Mendel wrote about himself (in the third person): "When he looked back on his own past, as a peasant lad in Heitzeindorf, who had had so hard a struggle to achieve a high-school education, often ailing and always poor ... he cannot but have been amazed to find himself at forty-six a mitred abbot."[14] But when having become the abbot, the guidance of the monks and the administration of the abbey put and end to what had been Brother Gregor's passion: the cultivation of pea plants and the search for their ways of hybridization. By that time he had planted and studied some 30,000 such delicately coloured and scented plants in the gardens of the monastery.

Mendel is often represented as a lonely monk with an obsession, tending his pea plants between four grim walls. In fact he was a learned man whom his monastery had sent to the University of Vienna, and he was for some time a teacher of physics. Far from being lonely and some-what abstracted, Brother Gregor was well-informed about the scientific literature in his branch of research, and he communicated with several prominent authorities about his findings in particular and biological topics in general. He read his paper, *Experiments in Plant Hybridization*, at two meetings of the Natural History Society of Brno, and it was published in *Proceedings of the Natural History Society* of the same town in 1866.

Mendel's laws are still the fundamental laws of heredity today. They say that if you hybridize, or cross, or mix two different plants, "one in four plants of their offspring has purebred recessive alleles, two out of four are hybrid, and one out of four is purebred dominant."[15] In simpler terms: if you cross a red plant with a yellow one, in the offspring you will find one red, one yellow, and two orange. This is a *statistical* statement that says nothing about the way this happens, about the hereditary mechanism. For "Mendel [like Darwin] knew nothing about the mechanism of fertilization, nor about genes, nor did he know the rudimentary facts of genetics."[16] The reason of this ignorance was quite simple: none of this knowledge was yet available. The process by which a sperm cell fertilizes an egg cell was discovered in 1875; in the same year the division of a plant cell and chromosomes were observed; and around 1875 was found that there is the same number of chromo-somes in the egg and the spermatozoon. Genes, mutations and the cell

---

14. Edward Larson: op. cit., p. 162.
15. http://en.wikipedia.org/wiki/gregor_mendel
16. André Pichot: *Histoire de la notion de vie,* p. 858.

components would become known many years later, and the history of genetic discoveries is far from over at present.

Why are Mendel's laws so important? They were the first experimental explanation of the way in which plants and animals descend from each other. Up to Mendel biologists supposed that offspring was simply a mixture of the characteristics of both parents.

In a child you would find half the characteristics of the father and half of the mother. Close observation showed that this was not true, and the closest observers were evidently people who worked with plants and animals day-in and day-out. The controversies about evolution had sharpened the interest of many gardeners and breeders, and this resulted in a spate of hereditary theories.

One of these theories was Darwin's idea of "pangenesis" and the "gemmules," which we have encountered earlier. He proposed that every organ and part of an organ secreted a kind of replica which was carried via the blood stream to the sex organs, through which it was communicated to the offspring. Without the slightest experimental justification for this process, this idea illustrates the extremes to which his guesswork could go. Even though Mendel published his findings in 1866, six years before the publication of *The Descent of Man*, Darwin never knew about them and therefore did not take them into account in his revisions of the *Origin*. Mendel's work will remain practically unknown till its rediscovery around the turn of the century. Up to then it was just another among the many theories presented in all sorts of learned meetings and publications by "gardener-crossbreeders."

Cutting the Tails of Mice

"What at present we consider to be Darwinism should in fact be called Weismannism," writes André Pichot. "Weismann is the author of the first coherent and complete theory of heredity."[17] This is high praise for a man whose name is barely remembered, except by the experts. Yet there is no gainsaying that the German evolutionary theorist August Weismann (1834-1914) has played a leading role in the study of the evolution of life on our planet. "With Weismann we really witness the birth of 'Darwinism', through the formulation of a theory of heredity."[18]

---

17. André Pichot: op. cit., pp. 860, 862.
18. id., p. 906.

Darwin's sketch of an evolutionary theory had remained a kind of hollow construction because it lacked the essential filling of, precisely, a scientific theory of heredity. Darwin improvised his explication of "pangenesis" and "gemmules," but nobody took it seriously because it was so patently spun by fancy. Weismann, a physician by training, became interested in the problem of evolution, which was then *the* topic among the intelligentsia thanks to Darwin's *Origin* and the propaganda by his proponents in Britain (the X Club, Francis Galton and Herbert Spencer), France (Clémence Royer) and Germany (Ernst Haeckel).

The Germany of those years was the most advanced industrial nation in the world, hard-working and of a dutiful accuracy in all its undertakings. It was the topmost chemical producer, and its machinery, especially its optical instruments, enjoyed worldwide renown. The discoveries of the composition of the cell and of genetics in general should be seen against this background; they were made mainly by German scientists and others who in one way or another were connected with German universities. The cell nucleus, mitosis, meiosis, chromosomes and gametes were all discovered and given their names in this period. It was the work of these cytologists that Weismann used to formulate his theoretical conjectures.

In 1889 he presented his "germ plasm theory" at the University of Göttingen. It states that "multicellular organisms consist of germ cells containing heritable information, and somatic cells that carry out ordinary bodily functions. The germ cells are influenced neither by environmental influences nor by learning or morphological changes that happen during the lifetime of an organism, which information is lost after each generation."[19] This meant that each multicellular organism consisted of two kinds of cells: the small mass of the reproductive cells and the huge mass of all the other cells. The reproductive cells, the "germ plasm," cannot be influenced or changed by anything from the rest of the organism or its environment. The reproductive cells will reproduce themselves in the reproductive acts of the organism, but the other cells of the organism will die and disintegrate into their molecular components.

Seen like this, there was an insurmountable separation between the germ cells and the rest of the body cells. This separation was called "the Weismann barrier" and is central to the Modern Evolutionary Synthesis. Weismann made evolution become a matter of the germ cells, more

---

19. http://en.wikipedia.org/wiki/august_weismann.

specifically of the small particles which he said they contained: the biophores ("carriers of life"). It should be stressed that those biophores were fictitious; neither they, their composition or their effects had been observed. Still they may be seen as the ancestors of the genes.

Weismann sharply opposed his theory against the inheritance of acquired characteristics. He supposed this theory to be Lamarck's, misled by the fact that the work of Jean-Baptiste Lamarck was becoming neglected, or heedful not to stain the fame of the idol of the evolutionists, Charles Darwin. Weismann was actually an anti-Darwinian on two accounts: Darwin had in the revised editions of the *Origin* integrated more and more Lamarckian elements, and the germ plasm with its biophores condemned the gemmules to oblivion. When Weismann proclaimed his theory, Mendel had not yet been rediscovered. Once this happened, one or two decades later, the original Darwinism would be hardly recognizable: 'Darwinism' was born – and Lamarck was burdened with a reputation which will turn him into the semi-grotesque antagonist of 'Darwinism' even in writings of present-day authors.

To prove the futility of the theory of acquired characteristics, Weismann cut the tails of 1500 mice (some say rats) belonging to 20 successive generations. He did this to show that the thesis that characteristics could be transferred from parents to their offspring was false. In the chapter on Lamarck we have seen that such transfers were (and are) not only a popular belief, but that this was a common conviction among all naturalists, including Darwin. Weismann also wanted to disprove the myth that some Jewish men were born without a foreskin, due to the millennia-old practice of circumcision. Like so many others he had a wrong understanding of Lamarck, thinking that the inheritance of characteristics was acquired through "effort or will." What Lamarck had said was that they were acquired through a *besoin*, a need of the organism, caused by the necessities of its living conditions, which created a habit that would result in the change of an organ or the growth of a new one.

Behind this historical facts something essential and dramatic unfolded: "life" disappeared from the biological sciences. The soft sciences of biology were generally estimated to be at least a century behind the hard physical sciences. In physics things could be measured, counted and calculated in order to make predictions. In biology things could also be counted, sometimes measured, but never calculated to make predictions. It was the ambition of the biological theorists to forge their discipline into a hard science too.

What could not be defined, measured, counted or calculated was the element of "life" in living things, that what made biology "the knowledge of life." An important aspect of the impulse driving Darwin had been to prove natural theology wrong and describe evolution in terms of a hard science. He had wrestled with the "Creator," and grudgingly allowed room for him in the revisions of his *Origin*. Weismann's first publication on evolution had been a comparison between creationism and Darwinism. His biophores were still "carriers of life," but the hereditary process began now to be seen as a 'mechanism', and life would soon dry up among the lifeless components of the mechanism. We are approaching the year 1900, the turn of the century when occultism had a heyday, but in academic science "mysticism" became a funny word.

Mutations and Saltations

The third pillar of 'Darwinism', sometimes called neo-Darwinism, was, after Mendel and Weismann, the Dutch botanist Hugo de Vries (1848-1935). He was one of those foreigners who were trained at German science institutions, in his case the University of Heidelberg and the laboratory of Julius Sachs at Würzburg.

In his experiments of plant breeding, not with peas but with the evening primrose, de Vries had come to the same statistical conclusions as Gregor Johann Mendel. This drew his attention to Mendel's obscure paper of some thirty years earlier. When in 1900 he published his findings, though, he omitted to refer to Mendel's work, an 'oversight' which was found out by another botanist, Carl Correns. At the same time a third botanist, Erich von Tschermiak, blew the dust from Mendel's publication, with the result that this (still) essential contribution to the science of heredity was suddenly brought to light by three researchers at once.

De Vries' *Intracellular Pangenesis* appeared in 1889. The word 'pangenesis' reminds of Darwin, and it was indeed from Darwin that de Vries had his original idea. However, his theory differed from Darwin's in a way which will turn him eventually into an opponent of Darwinism. Doing away with the fictional "gemmules," he proposed that heredity depended on particles within the cells, intracellular, which he called "pangenes." The inherited characteristics were then no longer a fifty-fifty mixture of the characteristics of the parental progenitors, but depended, each of them, on some particle in the core of the cell. This is the origin of the idea that each characteristic of an organism is tied to a specific material element in the hereditary matter, now called a 'gene', which is an abbreviation of 'pangene'. De Vries' pangenes were no less theoretical than Darwin's gemmules; the confirmation of their existence and function will have to wait another half century.

Around 1900 a drastic turnabout in de Vries' theory of heredity ensued. In that year the original German edition of the first volume of his *Mutation Theory* was published. By "mutation" de Vries meant a sudden unexplainable change in one of the pangenes, resulting in a change of one of the phenotypic characteristics of the organism. (The 'genotype' is the hereditary material of an organism, the 'phenotype' the way this material is expressed in its physical make-up.) But what de Vries understood by his revolutionary notion of the mutation was quite different from what this concept, fundamental in biology, means now. He conceived a mutation as radical to such a degree that it did not contribute to or cause a phenotypic characteristic, but that it created a new species. "For him species do not change progressively, they appear directly in their finished form through sudden jumps, without any apparent reason, and he calls such jumps 'mutations'."[20]

According to Darwin, the whole process of evolution took place gradually, bit by bit, small modification by small modification. This law of his theory may be found on every other page of the *Origin*. For example: "Why should not Nature take a sudden step from structure to structure? On the theory of natural selection, we can clearly understand why she should not; for natural selection acts only by taking advantage of slight successive variations; she can never take a great and sudden leap, but must advance by short and sure, though slow steps."[21] Therefore de Vries' "theory of mutation has originally been conceived *against* Darwin's hypothesis of the gradual transformation of the species ... The mutation is an abrupt and sudden variation, a jump from one form to another, a discontinuity ... It is a replacement of Darwin's theory of gradual transformation ... One single mutation suffices to effect a change of species ..."[22]

Mutations were supposed to explain Darwin's variations, for which he had no explanation (de Vries could not explain mutations either). They set up de Vries' evolutionary theory as "a rival of Darwin's theory; in fact, they turn it into anti-Darwinism." The anti-Darwinian stance

---

20. André Pichot: op. cit., p. 907.
21. Charles Darwin: op. cit., p. 181.
22. André Pichot: op. cit., pp. 907 (emphasis added), 912, 914.

was enforced by the fact that the concept of the mutations led to a theory of saltationism, which holds that evolution has not happened gradually but by something like quantum jumps. Even Daniel Dennett recognizes that "the people early in [the twentieth] century who redis-covered Mendel at first thought of themselves as anti-Darwinians."[23] This may come as a surprise, for the mutation theory in its revised version is at present one of the essential concepts of 'Darwinism', and de Vries has been added to the pantheon of great Darwinians.

Seen from the viewpoint of the period around 1900, this accumula-tion of theories – for there were also the several versions of Lamarckism, saltationism (e.g. St George Mivart), orthogenesis, the ever resistant creationism, and others – resulted in "an astounding hodgepodge of biological theories," in the words of Pichot. Larson agrees: "No con-sensus existed among biologists about how evolution operated."[24] The scientific problems remained enormous: the age of the Earth (calcu-lated by Kelvin) was too short; the mechanism of inheritance remained unknown; the gaps in the fossil record persisted. "Lamarckism and orthogenesis seemed to solve too many problems to be dismissed out of hand," writes Larson, and they continue to ask questions of the standard theory. All the same, the fame of Darwinism, built up by the crusading activism of its proponents, had in academic circles augmented to a degree that "Darwin had become the obligatory reference" and all plausible theories, even anti-Darwinian, were in some way integrated into in what was in the course of being invented: 'Darwinism'.

"The Modern Synthesis"

Trying to overcome the unholy confusion in the field of evolution-ary theory, a new instrument was used: mathematics. This may seem a rather inappropriate instrument to grasp the phenomenon of life with its infinite variations, often alike but never exactly the same, and of which the evolution is a flowing, historical, one-time process. Yet, Men-del's statistical approach could be analyzed mathematically, and instead of taking single organisms as their object of study, the mathematical biologists turned their attention towards whole groups or populations. "A satisfactory theory of natural selection must be quantitative," wrote

---

23. John Brockman (ed.): *The Third Culture*, p. 72.
24. Edward Larson: op. cit., p. 222.

J.B.S. Haldane. "By 1932, Haldane could open his book *The Causes of Evolution* by mocking the allegedly popular refrain 'Darwinism is dead'. The eclipse of Darwinism that had spread across evolutionary thought during the preceding generation had passed, he asserted."[25] All the same, if it had had to pass, it had certainly been there.

Not only was this therefore an effort to overcome the deadlock in the theoretical interpretation of the evolution of life on Earth, essential to it was also an eagerness of biology to match the physical sciences in the accuracy of their measurements and calculations. The time that the various subjects of biology were the study of amateurs lay, as shown in the previous chapters, not far in the past. Besides, the haziness of much of the acquired knowledge, be it in evolution, anthropology, paleontology, genetics, or whichever of the life sciences, stamped them as amateurish and 'soft' in the eyes of the practitioners of the 'hard' sciences. Any theory of evolution remained defective as long as it could not be based on an explanation of the way in which the characteristics of one organism were transmitted to an other, in other words on a theory of inheritance, of genetics.

Considerable progress had been made in the study of the unit of all life-forms, the cell, and its components, the organelles. It had been discovered that the secret of inheritance was to be sought in the nucleus, more specifically in its minuscule, threadlike components which a German scientist had called 'chromosomes' because, when preparing them for observation under a microscope, they were so easy to colour ("chromos" means colour). Still, at the time Thomas Morgan (1866-1945) worked towards a theory of inheritance, he knew nothing of the nature or the functions of the gene. He was quite aware of this, as he was of the dangers of speculation. The terms he used at first, in *The Mechanism of Mendelian Heredity* (1915) were the neutral words "factor" or "unit" within the framework of Mendel's system.

If *Drosophila melanogaster,* the humble fruit fly, has become one of the star performers in genetics, it is mainly because of the work of Morgan, whose team bred them by the millions in his laboratory. This kind of fly was well-suited to study: it had only four pairs of chromosomes, was extremely prolific so that the generations succeeded each other rapidly, and it was quite susceptible to mutations. All kinds of tiny monsters were produced, flies with white eyes instead of black, with four wings instead of two, with legs on their heads instead of antennae ... Thanks

---

25. id., pp. 223, 224.

to the fruit fly the intricacies of the statistical laws of heredity were deciphered. Not the gene yet, however. The gene became "a certain location on a chromosome which corresponded to a phenotypic character, and in this location there had to be an undefined 'something' causing the character to be in a certain way ..."[26]

Statistics can only be applied to the perceptible appearance of organisms, their behavior, their populations – to their 'phenotypes'. What remained of at least as great an importance was what actually happened inside the cell, in its nucleus, and in the components of which both consisted, and this had to do with the 'genotype', the complexities related to the elusive gene. Here were clearly two separate subjects of study which, though occasionally overlapping, moved on different terrains. For the time being the gene was supposed to be the part of a chromosome which caused a certain characteristic of the phenotype to be produced. This was supposed to be effected by the newly discovered enzymes, which were proteins. 'Enzyme' became the buzzword and the explanation of all wonders, great and small.

From this rudimentary phase in genetics has sprung the idea, still commonly accepted, that there is a gene that causes alcoholism, superintelligence or its opposite, anger, unusual sexual potency or its opposite, religion, and all sorts of illnesses or malformations. This will lead to the idea that in the constitution of an organism the genes are of prime importance, for they decide what the organism will consist of and look like. The genotypic level dominates the phenotypic level. We – you and me – are nothing but vehicles built by our genes, who have as their sole aim to survive and multiply. But more of this later.

At last towards the end of the 1930s a theoretical structure was built, strong enough to withstand most criticism, by joining genetics with populations genetics, mainly based on mathematics, and tying both of them together in the cover of Darwinism, authentic or not. This composite theory was called: "the modern synthesis," or "the new synthesis," or "neo-Darwinism." Edward Larson calls it "the modern neo-Darwinian synthesis;" and the *Oxford Dictionary of Biology* declares it to be "the current theory of the process of evolution."

One understands the pride with which the birth of 'Darwinism' was saluted. The hollowness of the original Darwinian theory was at last stuffed with some substance, provided by the disciplines of genetics and mathematics. Several great scientists worked on this, among the best

---

26. André Pichot: *Histoire de la notion de gène*, p. 131.

known J.B.S. Haldane, Ronald Fisher and Sewall Wright; most of them were "primarily mathematicians who constructed statistical models." There was also Theodosius Dobzhansky, an evolutionary geneticist who had pioneered the experiments with fruit flies, and who formulated the thesis that natural selection took place through mutations in genes. (The structure of the gene molecules remained unknown.) His *Genetics and the Origin of Species* (1937) became one of the foundations of the New Synthesis. And there were also Ernst Mayr and George Simpson, researchers as well as popularizers of 'Darwinism' in its new formula of " genetics plus natural selection."

And there was Julian Huxley (1887-1975), who in his person integrated and symbolized this whole scientific upturn. It was his influential book *Evolution: The Modern Synthesis* (1942) that gave the movement its name. It was he, grandson of the famous Thomas Huxley and brother of Aldous, the writer, and Andrew, Nobel laureate for medicine, who became the founding director-general of UNESCO, starred in BBC programmes, supervised the production of nature films and won an Oscar for one of them. For a time the names Julian Huxley and evolution became more or less synonymous, and he was ranked as one of the five best brains of Britain. "Huxley believed in human progress with a religious zeal, and he wanted others to share his belief. For him, progress was embodied in the evolutionary process that gave us birth and would carry us ever upward if we let it."[27]

Evolution, in its new, neo-Darwinian look, was riding a very high wave indeed. About Huxley's *Evolution* has been written that this book "declared the triumphal advent of neo-Darwinism," and "triumphal" is the word for the scientific and cultural mood in which the movement was steeped. The veteran evolutionist Ernst Mayr himself wrote about two Darwinian revolutions, the first in 1859, the second being the advent of the "new synthesis". "The Darwinism accepted since the evolutionary synthesis is best simply called 'Darwinism', because in most crucial aspects it agrees with the original Darwinism of 1859."[28] Really? Is a Darwinism with Malthusianism and without any notion of the all-important genetics the same as a 'Darwinism' without Malthusianism and with genetics – not to mention the dizzying increase in the knowledge of all related sciences by which evolution was put in a transformed perspective? "Darwin must be distinguished from modern

---

27. Edward Larson: op. cit., p. 248.
28. Ernst Mayr: op. cit., p. 96.

Darwinism," writes Tim Lewens. "One of the primary justifications for examining Darwin's own views is precisely to expose the frequent mismatches between the Darwin who is invoked by today's biologists eager to defend their corner, and the Darwin who wrote the *Origin of Species* and *The Descent of Man*."[29]

All those prominent scientists aforementioned will nonetheless continue to label their new synthesis of the evolutionary doctrines 'Darwinism' and thereby perpetuate the distortion of the history of biology and the never-ending discussions about it. Even Stephen Gould, who in 1980 declared the New Synthesis effectively dead after launching the anti-Darwinian theory of "punctuated equilibrium," will shout his love of Darwin from the widely read pages of his writings and be dubbed "North America's Darwinist-in-chief."

Darwinism Triumphant?

A century after the publication of the *Origin* and a century and a half after Darwin's birth, DNA was identified as the hereditary material in the cells by Oswald Avery (in 1947), and the structure of the DNA molecule was unravelled by James Watson and Francis Crick (in 1953). "The late 1950s saw a celebration of Darwinism. With scientists all but agreed on how evolution operated, its study had gained standing in science." The scientists named in the previous paragraphs had all acceded to high positions in learning and scientific authority. "Then came the centennials of the Darwin-Wallace papers in 1958 and *The Origin of Species* in 1959. Books and articles on Darwinism marked these occasions for the public. Scientific associations commemorated them with conferences hailing the founders of population genetics and the modern synthesis. Among these major symposia, Haldane presided over an international conference in Singapore; Huxley gave a keynote address at the University of Chicago; and Fisher, Haldane, Huxley, Mayr and Simpson received specials medals at festivities in London."[30] (There is a similar wave of celebrations of 'Darwinism' in progress at the time of writing this book, the year 2009.)

This rise into prominence and fame of 'Darwinism' had the effect it has in most cases: popularization, unreflecting acceptance, distortion,

---

29. Tim Lewens: *Darwin*, p. 6.
30. Edward Larson: op. cit., p. 247.

bias, abstraction into irrationalism, dogmatization ... "By the widely celebrated centenary of Darwin's *Origin of Species* in 1959, the modern synthesis had become virtual dogma within biology and its leading proponents sat atop the profession in chaired professorships at elite universities and on the boards of all the relevant societies."[31] The dictum of Theodosius Dobzahnsky became its slogan: "Nothing in biology makes sense except in the light of evolution," once more identifying the fact of evolution in general with the theory of neo-Darwinism in particular. Paul Davies words the same idea: "Darwinism is the central principle around which our understanding of biology is constructed," in fact meaning that evolution, unlabeled, is the central principle. The difference is pivotal, has consequences in all fields of human endeavour, and is the cause of a host of unending, vicious squabbles.

But is the science of evolution *that* monolithic, even in recent times? The late Stephen Gould was one of the most articulated voices in evolutionary biology, and the many volumes of his articles and essays are still on the shelves of bookshops everywhere. Being closely involved in the problem, he was quite aware of it and wrote: "We [evolutionists] have always acknowledged how far we are from completely understanding the mechanisms (theory) by which evolution (fact) occurred. Darwin continually emphasized the difference between his two great and separate accomplishments: establishing the fact of evolution, and proposing a theory – natural selection – to explain the mechanism of evolution."[32]

From a recent vantage point Steven Rose said in an interview: "Alas, poor Darwin, more idle speculation and dogmatic assertion have been published in your name in the past two decades than in the full preceding century, and still the torrent continues."[33] But, as "criticizing Darwin is extremely unpopular among English-speaking biologists" (Jonathan Wells), one or two influential French voices may also be heard, e.g. Pierre Sonigo: "More than 150 years after having been formulated, the Darwinian theory of evolution is accepted but still not perfectly understood."[34] Or Gerard Amzallag: "Darwinism [i.e. neo-Darwinism] is not a scientific but a metaphysical theory."[35] While Jean Swyngedauw states forcefully: "To attribute the inventions of nature to the chance action of viable mutations, which is the basic principle of all forms of

---

31. Edward Larson: op. cit., p. 237.
32. Stephen Jay Gould: "Evolution as Fact and Theory" in *Discover*, May 1981.
33. John Horgan: *The Undiscovered Mind*, p. 187.
34. Pierre Sonigo and Isabelle Stengers: *L'Evolution*, p. 48.
35. Gerard Amzallag: *La raison malmenée*, p. 162.

Darwinism, is a gross barbarism unworthy of the biological sciences."[36] All these commentators, putting 'Darwinism' into question, are experts in their fields of biology.

So closely are the sciences of life connected with the meaning we give to life that any statement about the basics of evolution acquires an ideological or metaphysical aura. Darwinism has since its beginnings carried atheism, not to say anti-theism, in its banner, and ostentatiously so. This is totally understandable after reading in the *Origin*: "I formerly spoke to very many naturalists on the subject of evolution, and never once met with any sympathetic agreement,"[37] and realizing that professing a naturalistic theory in the open meant confronting the near-totality of the established order. As a result the stance of the biological sciences against religion in all its aspects is much more outspoken than that of the physical sciences. A mathematical formula can be considered neutral; a statement about life is often implicitly judgmental, philosophical, metaphysical.

Yet it is difficult to gainsay that some biological scientists are revelling in the authority they think their knowledge confers upon them. In this spirit Carl Zimmer writes about "taking Darwin's cold bath," by which he means accepting the pointless view of life which is the reality propagated by Darwinian evolution. "By taking the Darwinian 'cold bath', and staring a factual reality in the face, we can finally abandon the cardinal false hope of the ages – that factual nature can specify the meaning of our life by validating our inherent superiority, or by proving that evolution exists to generate us as the summit of life's purpose."[38] Some materialistic-minded persons-of-science seem to find a delight without end in proclaiming that the human being is no longer made in the image of God, but that he is "an accidental and incidental product of the material development of the universe, almost wholly irrelevant and readily ignored in any general description of its functioning." (Scott Atran[39])

"Darwinism espoused an ultimate materialism," writes Larry Witham. "Even the modern synthesis, adding the variable of genes to natural selection, viewed chemical mutations as being shaped by chance forces in the environment."[40] It is against these "chance forces"

---

36. Jean Swyngedauw: *A l'origine de la vie le hasard*, p. 131.
37. Charles Darwin: op. cit., p. 445.
38. Carl Zimmer: *Evolution*, p. xvi.
39. John Brockman (ed.): *What is Your Dangerous Idea?* p. 173.
40. Larry Witham: *The Measure of God*, p. 242.

effecting extremely complex and meaningful results that Jean Swyngedauw protested a few paragraphs ago. Others are at least as forceful in their reaction, e.g. David Berlinski: "Darwinism is no longer merely a scientific theory but an ideology. ... The term 'Darwinism' conveys the suggestion of a secular ideology, a global system of belief. So it does, and so it surely is. ... Darwinism has achieved the status of inviolable science, combining the dogmatism of religion with the entitlement of science."[41] Arthur Koestler signalled already half a century ago that the sciences of life had built a "citadel of orthodoxy." To them "the only scientific method worth that name is quantitative measurement; and, consequently, complex phenomena must be reduced to simple elements accessible to such treatment, without undue worry whether the specific characteristics of a complex phenomenon, for instance man, may be lost in the process."[42]

Julian Huxley and George Simpson "saw belief in God as a remnant of a prior stage in human psychological evolution. Like the human appendix, it no longer served an adaptive purpose and instead could harm." And Edward Wilson, the founder of sociobiology, "described religion (or at least ethics based on religion) as 'an illusion fobbed off on us by our genes to get us to cooperate'." Still, according to Larson, acceptance of the modern synthesis coexisted with all manner of religious faith. "Fisher clung to his Anglican heritage, while Wright tilted towards Protestant process theology and Haldane warmed to Hinduism. Dobzhansky remained a professing Orthodox Christian and, during the 1950s, embraced the efforts of Teilhard de Chardin ... Lack converted from agnosticism to evangelical Protestantism in the very year he published *Darwin's Finches*. Like many mainstream Catholics and Protestants, he accepted evolution to a point, but believed that God created human souls."[43]

What had, almost unnoticed, slipped through the fingers of the scientists, or through the intricate network of neurons in their brains, was the subject of all *biology*: Life. Weismann's "biophores" and de Vries' "pangenes" were material particles, but still carriers of life. "Because of the progress in biochemistry, the theory of 'elementary living particles', at the origin of the biophores and pangenes, fell in disuse. There is no clean and abrupt break, but mentions of those elementary living

41. William Dembski (ed.): *Uncommon Dissent*, pp. xxxii, xxxiii.
42. Arthur Koestler: *The Ghost in the Machine*, p. 4.
43. Edward Larson: op. cit., p. 250, 274, 250.

particles have little by little disappeared."[44] The modern synthesis will accept nothing but "a purely materialistic chemical process". And so it has been ever since.

The problem of course is: how does one measure life, omnipresent but invisible and ungraspable? How does one even define life? If science is materialistic, it is because to us, beings incarnated in matter, only matter is directly perceptible, and only experiments with material objects are communicable and repeatable. This is the starting point of the last four centuries in the history of humanity, when science has been such a powerful support for the realizations of the technology which shaped an amazing new world.

However – and this is a resounding 'however' – that life differs from matter, and that organisms cannot live without this different element, is also a fact. A dead bird does not move, a living bird flies away. The fault is not with materialism as such, as a practicable theory invented with great pains and propagated with great risk: materialistic science is one of the great achievements of humankind. At fault is the fact that this materialism has been declared the exclusive metaphysical basis of the understanding of anything whatsoever. Materialism is a partial approach of Reality of which Life and Mind (or Consciousness) are equally essential elements. This is the reason that vitalism and the countless unorthodox mental interpretations of our world keep cropping up again and again. Our knowledge is incomplete. The knowledge of our world and ourselves is incomplete.

"Perhaps the problem is that for some scientists reductionism [i.e. the method of scientific materialism] functions as a security blanket. It avoids the need to ask too many questions, to stare into the abyss of fundamental uncertainty. If we abandoned the universality of the reductionist approach, who knows what would happen? For sure, the nature of biological science would change. But so it should! This change is long overdue."[45] Thus writes Denis Noble, emeritus professor of Oxford University. In this opinion he is seconded by Jean Dorst, former director of the *Muséum d'histoire naturelle*: "Biology can agree with a conception which differs from materialism and reductionism. The latter are solutions of the easy way: it is so easy not having to ask oneself questions anymore. ... I would even say that what the science of our day has gained, is that it goes against the reductionist view of nature."[46]

---

44. André Pichot: *Histoire de la notion de gène.* 103.
45. Denis Noble: *The Music of Life – Biology beyond the Genome*, p. 66.
46. In *L'Homme face à la Science*, p. 166.

In 'evolution' there is room for matter and life and mind; in 'Darwinism' there is not. This is the reason why it has been controversial for the greatest part of its existence, and that it will remain controversial in spite of the superficial triumph it has been enjoying for some time. Among the people in disagreement there are fanatical hecklers, but there are also savvy experts, as knowledgeable in their field of expertise as their colleagues in the opposite camp. Jonathan Wells, embryologist and author of *Icons of Evolution*, has said recently: "I predict that within twenty years people will look back on the present and wonder how so many seemingly smart people could have believed in Darwinian evolution."[47] In *Darwin's Black Box*, Lynn Margulis, at present one of the most famous biologists in the world, is quoted as saying that history will ultimately judge neo-Darwinism as "a minor twentieth century religious sect within the sprawling religious persuasion of Anglo-Saxon biology."[48]

And there is for example the testimony of the esteemed evolutionary biologist Robert Shapiro: "Adherents of the best-known theory [neo-Darwinism] have not responded to increasing adverse evidence by questioning the validity of their beliefs, in the best scientific tradition; rather, they have chosen to hold it as a truth beyond question, thereby enshrining it as mythology. In response, many alternative explanations have introduced even greater elements of mythology, until finally science has been abandoned in substance, though retained in name."[49]

Evolution is important; the theories to explain it must necessarily be tentative at present; the fast advancing scientific knowledge, especially in biochemistry, will prove the existing dogmatic doctrines wrong, and readjust our understanding of matter, life and mind. In this way the biological sciences from one side, and the physical sciences from the other, advance towards the grail of all knowledge: Reality.

---

47. Denyse O'Leary: *By Design or by Chance?* p. 187.
48. Michael Behe: *Darwin's Black Box*, p. 26.
49. Robert Shapiro: *Origins*, p. 32.

8.

Social Darwinism

Biology was called upon to explain inequality, particularly by those who felt themselves destined for superiority.

ERIC HOBSBAWM

The Copernican Principle

In his long years of reflection, it had become patent to Charles Darwin that the human being was a product of nature just like any other organism, an animal among animals, even if possibly a special one. If so, Homo sapiens had to obey the same laws of nature, the chief one being the general competition for survival, as an individual within the species and as part of a species among rival species. The struggle for life was an inevitable, constant condition of life in all circumstances and at all moments. The law of life was *bellum omnium versus omnes*: the war of all against all.

"Darwin encourages us to study ourselves in the same way we would study any other species," writes Tim Lewens. "We should see the human capacities which have fascinated philosophers as the products of evolutionary processes, capacities whose functions have been modified over time, and which still bear the marks of earlier roles. These capacities have shaped our physical, biological and social environments, and they have been shaped by those environments."[1]

David Barash puts it in plainer language: "Plants that commit rape and bacteria that spoil food are following evolutionary strategies that maximize their fitness. And, clearly, in neither case do the actors know what they are doing, or why. We human beings like to think that we are different. ... The world of reproducing beings seems ridden with selfishness and conflict. It is a world of individualists [i.e. egoists], each

---

1. Tim Lewens: *Darwin*, p. 242.

wanting to get ahead. It is a world in which everyone is set apart – mate from mate, brother from brother, parents from children. It is a world in which even love is a strategy and each of us is very, very much alone. But, like it or not, it is our world, created by the same genes that created us and all the rest of life."[2] As Desmond Morris, author of The Naked Ape, tells us, we are "just as much an animal as any other species ... The forest ape that became a ground ape that became a hunting ape that became a territorial ape has become a cultural ape."[3]

The pleasure positivist scientists draw from debasing the human being – they see this as putting it in its right place – is intriguing, if only because all those learned persons, being human, are appraising themselves also. This revised evaluation of the human is called 'the Copernican principle' or 'the principle of mediocrity', formulated by Stephen Hawking as follows: "We are such insignificant creatures on a minor planet of a very average star in the outer suburbs of one of a hundred billion galaxies. So it is difficult to believe in a God that would care about us or even notice our existence."

The fact that the humans in all parts of the world have up to now ranked themselves as the highest kind of being, as the lords and leaders of creation, is branded by Tim Lewens as "hubris", a Greek word meaning 'excessive pride'. Lewens writes: "Darwin's view also exposes, in the eyes of many, a kind of hubris that our species has been prone to. We are not distinct from nature, we do not ride above it, and neither are we evolution's greatest work ... Darwin is sometimes portrayed as one of a series of revolutionary thinkers who have exposed the modesty of man's position. Copernicus demonstrated that the Earth is not at the centre of the universe, but merely one of many planets revolving around the Sun. Darwin shows that Man is not a species apart from nature or above it, but, like all species, one among many of the branches of the tree of life. Darwin shows us that our self-image needs cutting down to size. Our species is unique, but uniqueness is ubiquitous in nature."[4]

---

2. David Barash: *Sociobiology – The Whisperings within*, pp. 31, 131.
3. Desmond Morris: *The Human Zoo*, p. 64, and *The Naked Ape*, p. 21.
4. Tim Lewens: op. cit., pp. 244-5.

Social Darwinism

Simultaneously the opinion was abroad of the superiority of the white 'race'. This racist, quasi-religious feeling of superiority, associated with the conviction that the task of the white Europeans consisted in civilizing the world, was at the time an inherent factor of European culture. "What at present we find abominably racist in the texts of Gobineau, Vacher de Lapouge, Darwin, Haeckel, Buchner, Vogt, Gumplowicz, and others, was at the time the dominant opinion, so commonplace that hardly anybody thought of criticizing it, neither on the left or on the right."[5]

"More to the point, Darwin was happy to describe races in terms of higher and lower, and he had no qualms about likening the lower human races to the higher apes. He believed that the whites were the highest of all races." In The Descent of Man, he wrote: "At some future period, not very distant as measured by centuries, the civilized races of man will almost certainly exterminate, and replace, the savage races throughout the world. At the same time the anthropomorphous apes will no doubt be exterminated. The break between man and his nearest allies will then be wider, for it will intervene between man in a more civilized state, as we may hope, even than the Caucasian, and some ape as low as the baboon, instead of now between the negro or Australian and the gorilla."[6]

Racial superiority complexes are generally associated with the horrors of Nazism and other kinds of strutting fascism. It is little known that at the time the racially most pretentious were the Anglo-Saxons. "The superiority of the Indo-Europeans – nowadays one tends to forget this – was not only a characteristic of the blond Germans, but also of the Anglo-Saxons. At the time there was even a veritable fashion of Anglo-Saxon superiority." (Pichot[7]) In the genealogical tree of the Indo-Germanic races as drawn by Ernst Haeckel during Darwin's lifetime,[8] the Anglo-Saxons occupy the top, on the same level as the High Germans, which is rather surprising, for the famous Herr Professor Haeckel was a super-nationalistic Pan-German. That some of the Anglo-Saxons were then at the peak of their colonial expansion, might be kept in mind. (And all this explains Charles Darwin's satisfaction

5. André Pichot: *La société pure de Darwin à Hitler*, p. 386.
6. Tim Lewens: op. cit., pp. 214-5.
7. André Pichot: op. cit., p. 69.
8. id., p. 331.

when the Beagle dropped anchor in Sidney: "My first feeling was to congratulate myself that I was born an Englishman.")

"By formulating the principle of the struggle for life and of natural selection," writes Pichot, "Darwin has not only revolutionized biology and natural philosophy, he has also transformed political science. ... The idea of applying Darwinism to the human society and politics has been immediate."[9] This goes to show how sudden was the impact of the Darwinian revolution, and how Darwin had brought into the open ideas which were ripening in the back of many minds. In the struggle for life, or for superiority of any kind (political, national, economical, cultural, racial) it was the natural law and therefore the inborn right of the best and brightest to win and to lord it over the others. The principle was simple, the urge generally innate, the execution in the present and the future as imperative as it had been in the past.

Herbert Spencer (1820-1903) led as it were a parallel life to that of Charles Darwin. He is mostly remembered for his pithy formula of the "struggle for life", inserted by Darwin into the title of his Origin. Spencer was nonetheless a great synthetic thinker who, all by himself, tried to give thought to what the nineteenth British century, the Victorian Age, represented. "The basis for Spencer's appeal to many of his generation was that he appeared to offer a ready-made system of belief which could substitute for conventional religious faith at a time when orthodox creeds were crumbling under the advances of modern science. Spencer's philosophical system seemed to demonstrate that it was possible to believe in the ultimate perfection of humanity on the basis of advanced scientific conceptions such as the first law of thermodynamics and biological evolution."[10]

The nineteenth century in western Europe was synonymous with the progress of humanity, and Spencer was called "the great exponent of Victorian optimism." He was, although now overlooked, one of those figures who dedicated their life to the advent of a new era, and his voluminous writings contributed to composing the new self-view of humanity, gradually replacing the traditional Christian view. No wonder that he was accepted as a member of Huxley's "X Club". His First Principles of a New System of Philosophy appeared in 1862. "Spencer posited that all structures in the universe develop from a simple, undifferentiated homogeneity to a complex, differentiated heterogeneity, while being

---

9. André Pichot: op. cit., p. 70.
10. http://en.wikipedia.org/wiki/herbert_spencer

accompanied by a process of greater integration of the differentiated parts."[11] This sounds Lamarckian, and, indeed, Spencer would basically remain a follower of Lamarck all his life, while grudgingly integrating natural selection into his worldview. Thus do the intricacies of human history combine in weaving the tapestries leading to the future.

The earliest and chief promoter of social Darwinism in France was the translator of *The Origin of Species*, Clémence Royer. In the extensive preface to her translation, published in 1862, she predicted that Darwin's discovery would without any doubt be more important from the social standpoint that from the biological, and she launched into a vehement diatribe against the Christian religion and democracy. "The law of natural selection, applied to humanity, makes us perceive, with amazement and pain, how wrong our political and civil laws have been up to now, as has been our religious morality." And she goes on to defend eugenics, favouring the strong and suppressing the weak. Georges Vacher de Lapouge, despite being a Frenchman, had a profound influence on the German racist thought and literature of the time. He wrote: "As soon as *The Origin of Species* was published, the clear-thinking minds understood that not only those ideas about history and the evolution of our societies, but the foundations of morality and politics could no longer remain what they had been until then."[12]

Ernst Haeckel (1834-1919) played in Germany a role comparable in many aspects to that of Spencer in Great Britain (and Auguste Comte in France), strengthened by his stature as a university professor. He was the main propagator of Darwinism, which he interpreted as a framework for a universal philosophy based on science. Just like Comte and Spencer before him, he projected a vision which resembled a religion. He proposed to fuse the inorganic with the organic and bring them under one law, but both were still in the rudimentary phases of scientific exploration. (The cell was supposed to be nothing but a blob of plasm, and the existence of the atom was still controversial.) Science as the a-religious religion of human progress was the keynote of the nineteenth century. It would lead to the perfect society of perfected human beings, 'supermen' in one of several variations imagined and longed-for at the time, of which Nietzsche's is the best known.[13]

Although the ideas of these once prominent people have been

11. http://en.wikipedia.org/wiki/herbert_spencer
12. André Pichot: op. cit., p. 142, 129.
13. See Georges Van Vrekhem: *Hitler and his God – The Background to the Nazi Phenomenon*, chapter 7 "Superior People" and chapter 8 "Long Skulls and Broad Skulls."

overtaken by the enormous expansion of the sciences, and have faded into the background of our awareness together with their authors, their contributions to the foundations of our world are considerable. Comte's positivism is still alive in contemporary thought; Spencer defined the social sciences; Royer helped detaching our world from the medieval world; and Haeckel is still discussed for his 'recapitulation theory' which says that the growth of an organism (ontogeny) repeats the evolutionary development of its kind or species (phylogeny). But, as mentioned before, Haeckel was "a polemical German-nationalistic chauvinist," for whose attitude the following quotation from the beginning of the Great War may suffice: "One fully-trained German warrior has a higher intellectual and moral life-value than hundreds of the crude nature-people whom England, France, Russia and Italy put up against him."[14]

Practicing Social Darwinism

Social Darwinism set the tone in the worsening tensions between the European nations in the last decades of the nineteenth century and at the beginning of the twentieth. Haeckel's voice was but a typical one among many in every sizable nation. Darwinism equipped each government to whatever it claimed – territory, markets, science, technical innovation, cultural superiority – with a 'scientific' justification. For the struggle for life was the law of the world; the victor proved he was the best and therefore entitled to the spoils. "All the warmongers of that time – and there were many – made Darwinism and its social applications their own ... One should recall how much, and in what ways, Darwinism had an impact on the biology at the end of the nineteenth century. Struggle, rivalry and natural superiority had become universal explanations, and this in all domains."[15]

The following quotes belong to a mentality from a former era – except when one translates them into a present context. (For, as Vacher de Lapouge asked,

how can the world change if humanity does not?) In his Principles of Sociology, Herbert Spencer wrote: "We have to recognize that the struggle for life among the societies has been the instrument of their evolution. ... The basis of social cooperation is the combined effort

---

14. http://de.wikipedia.org/wiki/ernst_haeckel
15. André Pichot: op. cit., pp. 59, 61.

for attack and defense; it is from this kind of cooperation that all other kinds ensue. It is no doubt impossible to legitimize the horrors caused by this universal antagonism which, starting with the chronic wars between small groups ten thousand years ago, has led to the big battles between great nations. However, one must recognize that, without these horrors, the world would still only be inhabited by men of the weaker type, seeking shelter in caves and living from raw food. ... The inter-social struggle for life has been an indispensable condition for the evolution of the societies. ... We recognize that we are indebted to war for the formation of the great societies and the development of their organizations."

Ernest Renan, the great historian and expert in near-Eastern cultures, especially known for his *Vie de Jésus* (1863, life of Jesus), was also a social Darwinian, to which testify the following words of his. "If the foolishness, the negligence, the laziness and the lack of foresight of the nations did not cause them to fight each other, if would be difficult to say to what degree of debasement the human species would descent. Seen in this way, war is one of the conditions of progress, the whiplash which prevents a nation to fall asleep by forcing the mediocre mass of its citizens, satisfied with themselves, to wake up from their apathy. The day humanity would become a big Roman empire, living in peace and not having any enemies at its borders anymore, would be the day when its morals and its intellect would run the greatest danger."

In conclusion, these are the words of an early critic of social Darwinism, Jacques Novicow, who wrote in 1910: "Darwinism has profited from the archaic instincts of brutality, so deeply ingrained in the brains of the traditionalists, the conformists and the ignoramuses of whom still consists, unfortunately, the immense majority of the human race. When the Darwinian theories came in fashion, Marshall von Moltke [the epitome of Prussian militarism] could write with a glimmer of scientific justification that war, in other words collective homicide, 'was in agreement with the order of things established by God', because this 'order established by God' corresponded to perfection with the formulation of the 'laws of nature' as used by the positivists and Darwinians." [16]

---

16. André Pichot: op. cit., pp. 55, 57, 74.

Breeding a Superman

Looking back on the amazing story of discoveries, controversies, search, dedication, endurance, obstinacy, intelligence and invention, we find that the impulse at the core of the biological sciences, as of all science, was the aspiration to find out the truth of things, the reality behind the appearances of which the world of the humans consists. The myth of creationism could no longer be upheld as infallible truth against reason and its discoveries in geology, paleontology, anatomy and the global exploratory excursions of the naturalists, all fitting within the intuition behind the chain of being and Linnaeus classification. These facts of life, in their turn, matched with the general experience of the law of life which was the law of the strongest, "the survival of the fittest," effective among all organisms and among the humans as far as memory could look back. The law of the strongest was ostensibly also the law of the best, of positive physical and mental characteristics running in families and races, and dominant of equally inheritable negative characteristics. Always there were strong and there were weak, noble and ignoble, winners and losers, conquerors and conquered, healthy and unhealthy, intelligent and dull, well-shaped and deformed. In some societies such categories and sub-categories were even fixed by laws allegedly issued by some Divine Authority.

Eugenics, the 'science' of breeding better animals and humans, was an application of social Darwinism, referable to Charles Darwin himself, who wrote in his *Descent of Man*: "The surgeon may harden himself when performing an operation, for he knows that he is acting for the good of his patient; but if we were intentionally to neglect the weak and helpless, it could only be for a contingent benefit, with an overwhelming present evil. We must therefore bear the undoubtedly bad effects of the weak surviving and propagating their kind; but there appears to be at least one check in steady action, namely that the weaker and inferior members of society do not marry so freely as the sound [a questionable statement]; and this check might be indefinitely increased by the weak in body and mind refraining from marriage, though this is more to be hoped for than expected."[17]

These words could have been written by a fully fledged eugenicist. Still Darwin's epigones maintain that he recognized the principle of eugenics but refrained from advocating its implementation. It is true

---

17. Tim Lewens: op. cit., p. 219 (emphasis added).

that such an equivocal attitude was typical for the inveterate hedger that Darwin was, shunning controversy and even more so when after 1859 the criticisms came in and he kept revising the *Origin of Species*. He proclaimed natural selection but included Lamarckism in his considerations, e.g. writing about "the effects of the increased use and disuse of parts, which I have always maintained to be highly important." ("Use and disuse of parts" is a signal concept of Lamarckism.) As to evolutionary progress, about which he had doubts, he also writes: "As natural selection works solely by and for the good of each being, all corporeal and mental endowments will tend to progress towards perfection."[18] He brought down the human being among the animals, but concluded the *Origin* with a much-quoted phrase about "the grandeur" of his view of life. And in spite of his enormous struggle to become an agnostic, he brought back a "Creator" in the subsequent editions of his major opus. One can fully understand his vacillating position in the trying historical circumstances; but the sophistry of many of his disciples, when buttressing their statements with quotes from Darwin's "nebulous" doctrine, is often annoying.

The pseudo-scientific doctrines of social Darwinism and eugenics have had dire consequences in real life. The former justified the attitudes of peoples and nations which would lead to the First World War; the latter was the 'science' which justified the maiming, sequestration and extermination of thousands and even millions of human beings. This, of course, was something Darwin could not so much as have imagined, and for which he was as little responsible as Friedrich Nietzsche was for Nazism. It is also important to correct the common opinion that all this started with him. In previous chapters we have seen that evolution was in the air, and that Darwinism became the flag under which evolution conquered the thought and self-view of humanity because so many different theories could assemble under it. Proof of this is that Alfred Wallace, basing himself on the same ideas, reached the same conclusions. But also swimming in the same intellectual waters were Lamarck, Erasmus Darwin, Robert Chambers, Herbert Spencer, Thomas Huxley, and many others now less well remembered.

The father of eugenics is generally considered to have been Francis Galton (1822-1911), a relative of Charles Darwin. He published a first sketch of his theory of "eugenics" in 1865 and a full elaboration of it in 1869, in a book titled *Hereditary Genius*. He coined the word itself from

---

18. Charles Darwin: *The Origin of Species*, p. 200.

the Greek 'eu-genès' which means 'well-born'. One of his definitions of eugenics was "the study of all agencies under human control which can improve or impair the racial quality of future generations."[19] This abstract, innocent sounding words need some clarification. In practice eugenics was much more than "study," it was an implementation, often by law, of measures supposedly based on science to uphold and promote the purity of the human race and to counteract anything that might contribute to its degradation. 'Positive eugenics' consisted in the encouragement of the reproduction of males and females of good stock; 'negative eugenics' was intended to prevent the reproduction of males and females who were thought to be of defective stock.

If this sounds suspiciously like the breeding of animals, the suspicion is to the point. Galton himself wrote: "A man's natural abilities are derived by inheritance, under exactly the same limitations as are the form and physical features of the whole organic world. Consequently, as it is easy to obtain by careful selection a permanent breed of dogs or horses gifted with peculiar powers of running, or of doing anything else, so it would be quite practicable to produce a highly-gifted race of men by judicious marriages during several consecutive generations."[20] Edwin Black, in his book *War on the Weak – Eugenics and America's Campaign to Create a Master Race*, writes that the advocates of eugenics were primarily plant and animal breeders (besides the biological scientists and medical doctors). He quotes some of their enunciations: "The principles of heredity are the same in man and hogs and sun-flowers. – Every race-horse, every straight-backed bull, every premium pig tells us what we can do and what we must do for man. – The result of suppressing the poorest and breeding from the best would be the same for men as for cattle and sheep."[21]

The problem lay in the fact that eugenics was not a real science. What was true for the study of evolution as a whole, was also true for its applications: social Darwinism and eugenics could not be a real science as long as the mechanism of heredity remained unknown. Degenerated traits – to be blocked by negative eugenic measures – were supposed to be alcoholism, sexual perversion, blindness, criminality, cretinism, and feeble-mindedness in its many aspects, for which the term "moron" became fashionable. It is obvious that these specifications were much more inspired by social and moralistic norms than by science.

---

19. Edwin Black: *War on the Weak*, p. 18.
20. http://en.wikipedia.org/wiki/eugenics
21. Edwin Black: op. cit., pp. 36, 39.

Stronger still than morals and social standards was the age-old instinct behind the pitiless eugenic customs and laws in ancient societies. The healthy, strong and harmoniously built (among the own people) have always been admired and favoured, as they still are in our own time. Plato made eugenics one of the pillars of his *Republic*. Infanticide of the physically defective and superfluous has always been a widespread practice, and in Sparta, the classical example of authoritarian states, an obligation. "The Twelve Tables of the Roman Law, established early in the formation of the Roman Republic, stated in the fourth table that deformed children must be put to death. In addition, patriarchs in Roman society were given the right to 'discard' infants at their discretion."[22] These examples are taken only from Western history in relatively recent times.

The aforementioned may have reminded the reader of Nazi Germany and its genocidal extermination of the 'degenerated', animal-like Jews and Gypsies. It should now be clear, however, that the cradle of the doctrine of eugenics was not Germany but Britain. Pichot writes that in the first decades of the twentieth century "most of the geneticists and biologists" (including all scientists we have encountered thus far) supported eugenics, to the point of turning it into a religion. Galton had already written "Eugenics has become a faith. It must become a religious belief." Julian Huxley will write in 1936: "Once all the consequences of evolutionary biology will be found out, eugenics will inevitably become part of the religion of the future." And Julian Huxley, president to be of UNESCO, "will write in 1941, at the moment that the Nazis in their experimental gas chambers killed the mentally ill while the whole world was watching, that eugenics was an integral part of the religion of the future." Pichot comments wryly: "At the time humanism was simply not what it is today."[23]

"Give Me Your Tired, Your Poor ..."

Volumes have been written about all the prominent scientists who were active eugenicists, also in France, where a militant and racist nationalism will find its voice during the years of the nation-splitting Dreyfus Affair, and lead to collaboration with Hitler under the Vichy

---

22. Wikipedia: op. cit.
23. André Pichot: op. cit., pp. 9, 10.

regime. Still, the following statement by Edwin Black will no doubt come as a surprise to many: "The Nazi principle of Nordic superiority was not hatched in the Third Reich but on Long Island decades earlier."[24] Long Island is part of New York State in the United States of America. "Eugenics was conceived at the onset of the twentieth century and implemented by America's wealthiest, most powerful and most learned men against the nation's most vulnerable and helpless. Eugenicists sought to methodically terminate all the racial and ethnic groups, and social classes, they disliked or feared. It was nothing less than America's legalized campaign to breed a super race – and not just any super race. Eugenicists wanted a purely Germanic and Nordic super race, enjoying biological dominion over all others."[25]

The formation of the eugenic thought in Britain must be seen against the background of the industrial revolution which caused a sudden increase of the plebeian population, Marx' 'proletariat', in the towns. The living conditions of these poor people were appalling; illness (tuberculosis), alcoholism and illiteracy were rife among them. Thus was created a divide between the well-to-do and healthy on the one hand, and the poor and 'degenerate' on the other, all ranked within the typical and distinct British framework of class consciousness, which one finds back in the eugenic thought. Ronald Fisher, for instance, the mathematical evolutionist, is characterized by Edward Larson as having had "a stunning facility for mathematics and a brooding preoccupation with breeding better Britons,"[26] at a time that the British Union Jack ruled the oceans and considerable parts of the continents.

The general situation in the USA was quite different. There the industrial revolution had only an indirect effect. The direct occasion of the eugenic movement in the USA were the "eighteen million refugees and opportunity-seeking immigrants who arrived between 1890 and 1920: German Lutherans, Irish Catholics, Russian Jews, Slavonic Orthodox." They were uprooted, poor, spoke incomprehensible tongues, had suffered, looked dirty, were ill. "They did not mix or melt; for the most part they remained insoluble." It was with them in mind that Charles Davenport, chief USA eugenicist, wrote: "Can we build a wall high enough around this country so as to keep out these cheaper races, or will it be a feeble dam leaving it to our descendants to abandon the country to the blacks, browns and yellows, and seek asylum in New

---

24. Edwin Black: op. cit., p. xviii.
25. id., p. 7.
26. Edward Larson: *Evolution*, p. 225.

Zealand?"[27] This tone and tune were quite different from the famous lines by Emma Lazarus: "Give me your tired, your poor, your huddled masses yearning to breathe free, the wretched refuse of your teeming shore," graven on a tablet in the pedestal of the Statue of Liberty.

White, Caucasian, Protestant, Nordic, long-skulled (brachycephalic), blonde, blue-eyed – America's first eugenic propagandists believed that Germanics and Nordics comprised the supreme race, and that was what they wanted to preserve and promote. Race mixing was considered race degeneration, "mongrelization," corruption necessarily ending in race suicide. From Oscar McCulloch's *Blood of a Nation* (1902) onwards the purity of the blood (again a rather unscientific idea) became sacred. Every drop of inferior blood in a person's veins made him descend lower and lower in the racial scale. "The vast fortunes of Carnegie, Rockefeller and Harriman financed unprecedented eugenic research and lobbying organizations that developed international reach. ... Eugenics rocketed through academia, becoming an institution virtually overnight. By 1914 some forty-four major institutions offered eugenic instruction. Within a decade that number would swell to hundreds. ... The state of Indiana became the first jurisdiction in the world to legislate forced sterilization of its mentally impaired patients, poorhouse residents and prisoners. The practice would crisscross the United States,"[28] and rapidly spread through the entire world.

For "Eugenics targeted all mankind, so of course its scope was global. American eugenic evangelists spawned similar movements and practices throughout Europe, Latin America and Asia. Forced sterilization laws and regiments took root on every continent. Each local American eugenic ordinance or statute was promoted internationally as yet another precedent to be emulated by the international movement. A tightly-knit network of mainstream medical and eugenical journals, international meetings and conferences [1912 London, 1921 and 1931 New York] kept the generals and soldiers of eugenics up to date and armed their nation's next legislative opportunity."[29]

The main target of the supremacist white Americans, far superior to all those strange and suspect people who came to live on what they considered their soil, were of course the dark-skinned people, the Negroes, now called Blacks or African-Americans. "The cross between a white man and an [American-] Indian is an Indian, the cross between

27. Edwin Black: op. cit., pp. 22, 17.
28. Edwin Black: op. cit., p. 41.
29. id., p. xvi.

a white man and a Negro is a Negro." American-Indians there were few and they lived mainly on reservations; African-American there were many and you found them not only in the South but more and more everywhere. Racists – the whole Western way of thinking was Eurocentric and to a high degree racist in those heydays of colonialism – placed the Negro not much higher than the primates in the tree of life; to eugenicists he was a nuisance and a threat, comparable to what the Jews were to the Germans.

Charles Davenport wrote as follows: "We have in this country the grave problem of the negro, a race whose mental development is, on the average, far below the average of the Caucasian. Is there a prospect that we may through the education of the individual produce an improved race so that we may hope at last that the negro mind shall be as teachable, as elastic, as original, and as fruitful as the Caucasian? Or must future generations, indefinitely, start from the same low plane and yield the same meager results? We do not know; we have no data. Prevailing 'opinion' says we must face the later alternative. If this were so, it would be better to export the black race at once."[30] (At one time Hitler gave his consent for negotiations to export the German Jews to the island of Madagascar.)

"Despite their virulent racism, the Anglo-Saxon Clubs claimed they harboured no ill will toward the Negroes," writes Black. "Why? Because now it was [not racism but] just science – eugenic science. The Anglo-Saxon Clubs could boast: 'That "one drop of negro blood makes the negro" is no longer a theory based on race pride or colour prejudice, but a logically induced, scientific fact.' … This was a powerful redefinition of eugenics in action."[31] It is touching to place all this within the perspective we have thanks to our hindsight. What is narrated here happened less than one hundred years ago. Eugenics reached its zenith in the world in the 1930s. How many of us have ever heard that "from 1907 [Indiana] till 1960 more than 100,000 Americans were sterilized in more than 30 states"?[32] (About the 6 million Jews and the half a million Roma we have heard, although doubt has been sown in the minds of many by fanatic Holocaust-deniers.) The March on Selma took place in 1965, two years after the assassination of President John Kennedy and at the time of the Apollo space programme. – At the time of writing the USA has a Black President, Barack Obama.

---

30. id., p. 38.
31. Edwin Black: op. cit., p. 168.
32. http://www.sntp.net/eugenics/eugenics_america.htm

Animal Lovers

If the leading role of the USA in the eugenics movement surprises, the connections between American capitalism and Nazism in this field will also be novel information. "The Carnegie Institution, through its Cold Spring Harbour complex [in Long Island], enthusiastically propagandized for the Nazi regime and even distributed anti-Semitic Nazi Party films to American high schools. ... And there were the links between the Rockefeller Foundation's massive financial grants and the German scientific establishment that began the eugenic programs that were finished by Mengele in Auschwitz." No, Henry Ford and Charles Lindbergh were not the only Hitler fans.

The feeling of superiority was deeply ingrained in the Germanic people, who considered that dominating the world was their birthright. As to eugenics, the following words of the famous Herr Professor Haeckel, from his book *Die Lebenswunder* (the marvel of life, 1904), is one example chosen from many: "The killing of newborn misshapen children can therefore not be considered to be murder, as is still the case in our modern law books. On the contrary, we must approve of it and see it as an effective and useful measure as well for those concerned as for society. ... Hundreds of thousands of incurably ill persons are artificially kept alive in our modern nations, without this being of any use for themselves or for the community."[33]

All what we have read before, and much more, has been literally put into practice in many countries, in spite of sometimes being controversial in some cases, but especially in Nazi Germany and its conquered territories, where eugenics was the stringent law. Hitler's infamous Nuremberg Laws of 1935 are a direct application of the teachings of the American eugenicists, with the only difference that to the American "degenerates" were now joined by the Jews and the Gypsies. Thousands of physically and mentally handicapped were killed in newly invented gas chambers and with other methods which afterwards will be used in the extermination camps.

Once again, the criminal and in this case often bestial acts of an army of perpetrators ("Hitler's willing executioners") were justified by science. The racist and eugenic writings of the German medical specialists and biologists were taught in the army, the youth movements and the schools, of course in addition to Hitler's version of 'Darwinism' from

---

33. http://de.wikipedia.org/wiki/ernst_haeckel

Mein Kampf. The following quotations are from three men involved in the execution of the eugenic and extermination policies. Arthur Ostermann (1932): "Science is now able to establish the hereditary future of a person with a total certainty." Eugen Fischer (1933): "At present we now enough about the study of human heredity, and with a sufficient certitude." Otmar von Verschuer (1933): The bases for the application of the genetic measures are sufficiently assured." The medical experiments by the SS doctors in their camps makes gruesome reading.

"Man has become great through perpetual struggle. In perpetual peace his greatness must decline," wrote Adolf Hitler in *Mein Kampf.* In a speech to 10,000 young army lieutenants, his new generation of fanatical supporters, he said: "The whole universe seems to be dominated by this thought alone: that an eternal selection takes place in which the stronger keeps the right to remain alive and the weaker succumbs." In one of his 'table talks' late at night he said to the people of his intimate circle: "War is always, war is everywhere. There is no beginning to it, there is no peace, ever. War is life. War is in every contest, war is the primeval state." And: "The Jew: he poisons the blood of others but preserves his own blood unadulterated."[34] And when Germany in the winter months of 1945 will be crushed by the armies of the Allies, closing in from the West and from the East, he will give 'the Nero Order' to totally destroy any place that had to be abandoned to the enemy. He will also command what was left of the civilian population to leave their homes and walk away in the snow before the invaders, because they had proved to be not superior people but weaklings, and were therefore unworthy of him.

As André Pichot remarked: humanism then was not what it is today. 'Humanism' should be human, and in the course of history, as well as in our own time, it has proved to be many things which are the exact opposite of what the dictionaries, the philosophers and the holy men tell us that is human. In any anthology of the inhumanity of humans these words of Heinrich Himmler, founder and commander of the SS, Hitler's Black Order of the Death's Head, should be included, lest we forget: "We will never be brutal or heartless when it isn't necessary. We Germans, the only people in the world with a decent attitude towards animals, will also have a decent attitude towards these human animals," by whom he meant the Jews.[35]

---

34. Georges Van Vrekhem: op. cit., pp. 218, 219, 230.
35. id., p. 210.

9.

Sociobiology

... If we must marvel, let it be at our own presumption in imagining for a moment that we understand the many complex contingencies on which the existence of our own species depends.

CHARLES DARWIN

Edward Wilson – The Rise of the Genes

In our story of evolution, its interpretations and the consequences of these interpretations on the way contemporary humans see themselves, we arrive now at the period which has devised the ideas in the books on the shelves of our favourite bookshop. These ideas shape our thoughts and become part of our lives. In essence, the contemporary ideas about evolution and the human species in it are the same that originated in Darwin's time and were later known as "social Darwinism." But the general attitude of biology has irrevocably moved away from anything connected with the non-material, and proclaims itself solidly founded on the rock of material reality.

Edward Wilson's *Sociobiology: The New Synthesis* was published in 1975. Like Darwin's *Origins* in 1859 and Julian Huxley's *Evolution: The Modern Synthesis* in 1942, Wilson's book was the catalysis of a thought movement in search of a formulation. In the mid-70s a spate of books appeared about aspects of the central theme, the evolutionary roots of the human as a social being. Titles were e.g. *Animal Behaviour: An Evolutionary Approach* (John Alcock), *The Evolution of Behaviour* (Jerram Brown), *Human Ethology* (Irenäus Eibl-Eibesfeldt) – all, fully or in part, about the behaviour of the human animal as fashioned by evolution.

Desmond Morris had shocked the world with the *The Naked Ape* in 1967; Richard Dawkins' *The Selfish Gene* will closely follow Wilson's central opus in 1976, and *Sociobiology: The Whisperings Within*, by David Barash will contribute to the spreading of the glad tidings in 1980: "We humans are self-conscious animals. ... Our builders are our genes. ... The *raison d'être* of genes is purely self-propagation ..."[1] And that we are no more than robots created by genes to accomplish their selfish ends.

Edward Wilson (born in 1929) was an entomologist who "modestly thought of himself as the world authority on social insects."[2] He worked in the footsteps of a trio of famous ethologists: Karl von Frisch, who deciphered the dance of the bees, Konrad Lorenz, famous for imprinting his goslings, and Nikolaas Tinbergen, the theorist of animal aggression. This trio shared the Nobel Prize in Physiology or Medicine in 1973. We know that along with the biological research in the laboratories and in the field there was also a branch of mathematical biology, originally based on the statistical method of Gregor Mendel. Mathematics applied of necessity to groups of organisms and flourished decades before the structure of the gene was discovered. But now it *was* discovered, in 1953 by James Watson and Francis Crick, "an achievement that stands unrivaled in the annals of twentieth-century biology."

Wilson, the student of bees, ants and wasps *in vivo*, resisted the abstract methods of the mathematicians. Till he read a seminal article by William Hamilton on kin selection: "The Genetical Evolution of Social Behaviour." This was the conversion which would lead to his theory of sociobiology. "I wanted to create a showcase for sociobiology using insects and, in so doing, demonstrate the organizing power of population biology. ... I had no intention of extending my studies beyond the social insects. ... [But then] I had a revelation. Vertebrates weren't different at all."[3] Humans, who belong to the phylum of the vertebrates, were no different from ants and wasps, for all of them were the products of evolution and therefore subject to the same laws.

The philosophy supporting this view was gross materialism: all is matter because there is not and cannot be anything but matter. The material animal that is the human had descended from other material animals and consisted of the same material stuff. Not only the body but also the instincts, feelings and consciousness were formed by evolution and accordingly material. Like other animals the human had developed social characteristics which were explainable only through evolution. Wilson studied social insects; he discovered a mathematical theory which seemed to explain the social comportment of his insects; he

---

1. David Barash: *Sociobiology: The Whisperings within*, pp. 41, 201, 91.
2. Edward Wilson: *Naturalist*, p. 320.
3. id., pp. 321, 322, 323.

undertook the huge step of applying his theory to all animals, including humans. Sociobiology was born.

As Wilson said: *"Homo sapiens* is after all a biological species." In the present-day context biology should not be associated with its literal meaning 'knowledge of life', for it is purely materialistic. "History did not begin 10,000 years ago in the villages of Anatolia and Jordan. It spans 2 million years of the life of the genus *Homo.* Deep history – by which I mean biological history – made us what we are, no less than culture. ... Let us now consider man in the free spirit of natural history, as though we were zoologists from another planet completing a catalog of social species on Earth. In this macroscopic view the humanities and social sciences shrink to specialized branches of biology; history, biography and fiction are the research protocols of human ethology; and anthropology and sociology constitute the sociobiology of a single primate species,"[4] to wit *Homo sapiens.* Zoology, ethology, anthropology, sociology – everything became part of a single encompassing field of knowledge: sociobiology.

"Sociobiology is defined as the systematic study of the biological basis of social behaviour and the organization of societies in all kind of organisms, including human beings," said Wilson.[5] Mary and John Gribbin condense this definition: "Sociobiology is the study of all forms of social behaviour, including humans." And they hammer it home again and again: "People are animals. ... The only way to get a deep understanding of human behaviour is to apply to humankind exactly those rules that have proved so successful in explaining the way other animals behave. ... We want to concentrate on animals, particularly mammals and specifically human beings."[6]

In the previous chapters we have followed the fading away of anything super- or para-natural, accompanied by a parallel hardening of the materialist, positivist, reductionist standpoint, sometimes called 'metaphysical materialism'. (This means a materialism that is not the consequence of experience, but a theoretical, *a priori* concept.) What is new here is the concept of the genes as the agents of evolution. In the history of science it is a common phenomenon that an important discovery is at once supposed to be determinant of everything else. (Examples are Newton's gravitation and Einstein's relativity.) This happened again when the structure of the DNA molecule, the double helix,

---

4. Edward Wilson: op. cit., pp. 328-9.
5. Ullica Segerstråle: *Defenders of the Truth,* p. 97.
6. Mary and John Gribbin: *Being Human,* pp.205, 207, 1, 217.

was found out. With this discovery started the enormous expansion of microbiology which became the leading discipline in biology. "Molecular genetics took off." In the ten years following the discovery of the structure of DNA were discovered the mechanism of DNA duplication, the role of messenger and transfer RNA, the genetic code, the mechanism of protein synthesis, and the general principles regulating this synthesis.

"By explaining life through chemistry, and the evolution of life up to the human species through Darwinism, the biological sciences claimed to complete the totality of all science as well as of its explanation of nature. As in the ideology of that time positivist knowledge constituted the ultimate goal of humanity, the new science, having become the all-round explication and the source of all truth, was supposed from then on to replace religion and morality," writes André Pichot.[7]

Indeed, the attitude behind sociobiology grew virulently atheistic, even anti-theistic and generally anti-religious. In this it shared the spirit of the times. In 1970 the biologist Jacques Monod published *Le hasard et la nécessité* (chance and necessity), which would cause a worldwide sensation. Monod, a Nobel Prize winner in 1965, literally preached scientific atheism in a style and spirit borrowed from French existentialism. To him religion in any form was "animism." He wrote: "All the animist systems have in various measures wanted to ignore, besmirch or diminish biological man, make people abhor him or be terrified by some traits of his animal condition. The ethics of knowledge, on the contrary, encourage man to respect and accept this heritage, knowing how to master it when necessary. ... Man knows at last that he is alone in the indifferent immensity of the universe where he has emerged by chance. ... Man has to awake finally from his age-long dream to be confronted with his total loneliness, his radical strangeness. He knows now that, like a Gypsy, his place is in the margin of the universe, and that he has to live there. This universe is deaf to his music and indifferent to his hopes, as it is to his sufferings or his crimes."[8]

Soon the physicist Steven Weinberg, another Nobel Prize winner, will write in *The First Three Minutes*: "The more the universe seems comprehensible, the more it also seems pointless. ... The effect to understand the universe is one of the very few things that lifts human life a little above the level of farce, and gives it some of the grace of

---

7. Mary and John Gribbin: op. cit., p. 212.
8. Jacues Monod: *Le hasard et la nécessité*, pp. 223, 225, 216.

tragedy."[9] Larry Witham quotes Weinberg as having said: "Whatever naturalism is, it is better than religion, which is tantamount to belief in fairies. ... I am all in favour of a dialogue between science and religion, but not a constructive dialogue." "Science has tended to destroy religion and has allowed intelligent people to reject God," asserts Weinberg, adding wryly: "We should not retreat from this accomplishment."[10] The chemist Peter Atkins, together with many others, joins the somber chorus: "We are the children of chaos, and the deep structure of change is decay. At root, there is only corruption, and the unstemmable tide of chaos. Gone is purpose; all that is left is direction. This is the bleakness we have to accept as we peer deeply and dispassionately into the heart of the Universe."[11]

Reinterpreting the World

"Basically, sociobiology is a radical doctrine," states David Barash. The word 'radical' has several shades of meaning. Which one is applicable in this case? Ullica Segerstråle writes that Edward Wilson showed great zeal in making sociobiology "a truly predictive science, encompassing all of social behaviour," and that his zeal was "closely tied to an old desire of his to prove the [Christian] theologians wrong."[12] These are revealing assertions. For "a truly predictive science" reminds us of the pretensions of the French mathematician and astronomer Pierre-Simon Laplace, who said that the future as well as the past of the entire universe could be known if all elements constituting the present were known. Wilson sought to establish mathematics as the basis of biology, and we find the reference to Laplace justified: "Wilson's moral aim was a *quantitative* explanation of all aspects of human social behaviour. He was also interested in being able to formulate a trajectory of mankind's future" as a substitute for divine prophecy.[13] Barash has the reflection: "It seems increasingly clear that sociobiology will cause us to revise our own self-image, just as the works of Copernicus, Darwin and Freud have done."[14]

9. Steven Weinberg: *The First Three Minutes*, p. 155.
10. Larry Witham: *By Design: Science and God*, p. 153.
11. Richard Dawkins: *Unweaving the Rainbow*, p. xi.
12. Ullica Segerstråle: op. cit., p. 38.
13. Ullica Segerstråle: op. cit., pp. 82, 40 (emphasis added).
14. David Barash: op. cit., p. 240.

"Wilson's larger goal was to explain religion," writes Segerstråle, and she refers to his "deadly ambition" (on a certain occasion acknowledged by himself: "Once again I was roused by the amphetamine of ambition"[15]). We find this attitude of his confirmed by the editors of *Alas Poor Darwin*: "Edward Wilson and other scientists have promoted the sociobiological model of human nature in popular books and magazines with missionary fervour, aiming to convert the unenlightened. … Their claims, their language and their style have striking religious overtones."[16]

There is no doubt that Wilson was convinced to have found the method of explaining and interpreting the world and everything in it, including the human species; he also evaluated his method, sociobiology, as surpassing any other explanation and interpretation, including those of the religions, and that sociobiology should therefore become the new religion of humanity. Monod, one of his precursors, had already written: "In order to survive, vitalism [to Monod a form of "animism," the chief scapegoat of his reductionism] needs that there remain in biology, if not real paradoxes, at least 'mysteries'. The developments of the last twenty years in molecular biology have considerably reduced the domain of those mysteries, leaving only the field of subjectivity, of 'consciousness', open to the speculations of the vitalists. One does not run a great risk in predicting that, in this still separate domain, the speculations of the vitalists will prove as sterile as anywhere else where they have been applied."

We keep in mind that, according to sociobiology, an offspring of social Darwinism, it is evolution which has produced everything we humans consist of. The active and determining elements of evolution are the genes. What we are – including our brains, of course, and therefore our consciousness – has been gradually developed in what preceded us in the evolution, more directly in the primates, the australopithecines and the various subspecies of *Homo*. Anything we are and do is subject to the Darwinian prime law that every organism "maximizes its fitness." Taking all this in consideration, and recalling the ideologies of race superiority and eugenics which were the fruits of social Darwinism, it is no wonder that the new ideas provoked an inimical reaction.

Sociobiology was accused of being racist, genetically deterministic, abolishing free will, robbing the human being of its dignity, sexist,

---

15. Edward Wilson: op. cit., p. 323.
16. Hilary and Steven Rose: *Alas Poor Darwin*, p. 15.

reactionary, and explaining injustice away.[17] The animosity against Edward Wilson, especially by American extreme leftists, reached the news desks in February 1978, after he had given a lecture at a symposium in Washington. A group of demonstrators had invaded the podium and "a young woman behind me picked up a pitcher of water and dumped the contents on my head." The audience had given him a standing ovation, writes Wilson, and he had proceeded with his lecture.[18] Yet, the sociobiology story was far from finished.

Richard Dawkins – The Triumph of the Genes

Richard Dawkins (born in 1941) is undeniably the most famous biologist alive. His books are sold in great numbers, he is a darling of the media, and he has a vocal following among scientists. Recently voted one of the world's three leading intellectuals, he is a superstar. Professionally Dawkins is an ethologist, i.e. expert in animal behaviour, trained at Oxford by Nikolaas Tinbergen, his "old maestro", and a student of the feeding habits of chickens. His formative years were marked by the upsurge of neo-Darwinism and the spectacular breakthrough of molecular biology.

Behind the scientific formation, however, one senses the character of an enthusiast who might easily turn into a zealot. "I care passionately about what is true and I never say anything that I do not believe to be right."[19] Hamilton's mathematical population genetics had for Dawkins "a visionary quality," but he found the way in which it was expressed "not full-throated enough." Adding this to the confession that "as an over-romantic graduate" he had been captivated by Pierre Teilhard de Chardin's *Phenomenon of Man* and the Omega Point, one will not be surprised by the tone of Dawkins' prose and the radicalism of his reasonings. This is one of several characteristics he has in common with Edward Wilson.

Another marked influence on him was scientism, more specifically the mentality of the nineteenth century reconditioned by the mind of Bertrand Russell, who wrote: "My view of religion is that of Lucretius. I regard it as a disease born of fear and as a source of untold misery to

17. David Barash: op. cit., pp. 232 ff.
18. Edward Wilson: op. cit., p. 149.
19. Richard Dawkins: *The Blind Watchmaker*, p. xiv.

the human race. ... Whatever knowledge is attainable, must be attained by scientific methods; and what science cannot discover, mankind cannot know."[20] Moreover Dawkins seems to have assimilated the thought of Jacques Monod, whose *Le hasard et la nécessité* (1970) became an international bestseller and whom he had heard lecture. At times Dawkins' words take existentialist Monodian overtones, for instance: "The universe that we observe has precisely the proportions we should expect if there is at bottom no design, no purpose, no evil and no good, nothing but blind, pitiless indifference,"[21] reminiscent of the human being pictured by Monod as a lonely Gypsy lost in the margin of the universe.

Dawkins has been called "the ultimate ultra-Darwinist." He sees himself as "an enthusiastic Darwinian ... a real-life Darwinian." Edward Wilson calls him "Darwin's most ardent representative on Earth."[22] Major Darwinian elements in Dawkins' thinking are, firstly, his stance against eighteenth century theological naturalism, of which William Paley had become the figurehead,[23] and which is the central theme of *The Blind Watchmaker.* Secondly, Dawkins sticks stubbornly to the theory of gradualism as a fundamental mechanism of evolution; in its defense he wrote another of his books, *Climbing Mount Improbable.* The agnosticism and atheism of Darwin's background might be taken as a third tie with the glorified master, but Darwin's hesitant and soul-rending inner struggle differed considerably from Dawkins' barnstorming anti-theism.

He sparked a furore with *The Selfish Gene,* the book that made "the gene's eye view" popular. In this way of explaining and evaluating a living organism the role of agent of life and its evolution, no longer being God, an animist force, or the phenotypal organism itself, is conferred on the genes, whose importance since the discovery of the double helix has taken on gigantic proportions. It is the genes which now determine all aspects of life, acting from a single, all-pervading motive: the selfish urge of self-replication and propagation. "From the selfish-gene point of view, we are robot survival machines," declares Dawkins, "and because genes themselves can't pick things up, catch things, eat things, or run

---

20. *Russell on Religion*, pp. 169, 107.
21. William Dembski (ed.): *Uncommon Dissent*, p. 101.
22. Edward Wilson: op. cit., p. 223.
23. See the chapters on Charles Darwin.

around, they have to do that by proxy, they have to build machines to do it for them. That is us. These machines are programmed in advance."[24]

Genes consist of DNA, deoxyribonucleic acid. The novel doctrine teaches that an organism is just DNA's way of making more DNA. "The argument of this book [*The Selfish Gene*] is that we, and all other animals, are machines created by our genes,"[25] states Dawkins as his chief premise. The animal is a robot. A human animal is a "lumbering robot," "a large vehicle or 'survival machine' built by a gene cooperative for the preservation of copies of each member of that cooperative."[26] Jacques Monod had already declared the cell to be a machine. Edward Wilson had written in his *Sociobiology*: "The individual organism is only the vehicle [of the genes], part of an elaborate device to preserve and spread them with the least possible biochemical perturbation." Machine, device, robot: life's riddle was finally solved – a few residual problems notwithstanding, for instance how the genes had originated, where they had their irresistible urge to replicate from, and how they managed to produce their astonishing replicating machinery, living nature.

"The selfish gene theory is Darwin's theory," affirms Dawkins in *The Selfish Gene*, "expressed in a way that Darwin did not choose but whose aptness, I should like to think, he would immediately have recognized and delighted in. It is in fact a logical outgrowth of orthodox neo-Darwinism."[27] (Elsewhere in the same book he writes: "Much of what Darwin said is, in detail, wrong. Darwin, if he read this book, would scarcely recognize his own original theory in it, though I hope he would like the way I put it."[28]) In the previous chapters has been shown how tentative and defective Darwin's original theory was, and how 'Darwinism' was 'invented' step by step by Weismann, de Vries, the mathematical biologists and the neo-Darwinists. If one finds any truth in this historical review, Dawkins affirmation must sound frenzied, and his admission "I wrote *The Selfish Gene* in something resembling a fever of excitement"[29] does not surprise.

There are more startling aspects to the Dawkins phenomenon. In *The Extended Phenotype*, "the book that, more than anything else I have achieved in my professional life, is my pride and joy," Dawkins goes

---

24. John Brockman (ed.): *The Third Culture*, p. 80.
25. Richard Dawkins: *The Selfish Gene*, p. viii.
26. id.: *The Blind Watchmaker*, p. 236.
27. id.: *The Selfish Gene*, p. x.
28. id., p. 195.
29. Richard Dawkins: op. cit., p. xii.

beyond the simple selfishness of the genes. He develops a theory which says that, in order to achieve their selfish aim of maximal replication, genes collaborate not only within the cell and the organism, but with other conglomerates of genes in other organisms and even in their physical environment. (The genotype, it may be remembered, is the genetic basis of an organism, the phenotype its concrete realization as a physical organism.) "Certainly in principle, and also in fact, the gene reaches out *through the individual body wall* and manipulates objects in the world outside, some of them inanimate, some of them other living beings, some of them a long way away. With only a little imagination we can see the gene as sitting at the centre of a radiating web of extended phenotypic power. And *an object in the world is the centre of a converging web of influences from many genes sitting in many organisms.* The long reach of the gene knows no obvious boundaries. The whole world is crisscrossed with causal arrows joining genes to phenotypic effects, far and near."[30]

No doubt, Dawkins is of the opinion that the genes are gifted with magic powers. Aware that the one-gene-one-effect doctrine became increasingly unsatisfactory, he switched more and more to groups, clusters or conglomerates of genes acting magically (in any case inexplicably) in unison to realize their common goal. But, according to scientific materialism, is there anything in the universe able to act *towards* something, can evolution be intentional, can matter be teleological? Of course not. Anyone who admits even a glimmer of such pre-scientific animism, supernaturalism or mysticism steps automatically outside the bounds of academic science. Therefore Dawkins declares sternly: "Selfish genes have no foresight. They are unconscious, blind replicators."[31] Which is not what his extended phenotype theory seems to mean or suggest.

And he goes further. He maintains that his view of the selfish gene and the extended phenotype "applies to living things everywhere in the universe," and thereby founds Universal Darwinism. "It is an established fact that all of life on this planet is shaped by Darwinian natural selection which also endows it with an overwhelming illusion of 'design'. I believe, but cannot prove, that the same is true all over the universe, wherever life may exist. I believe that all intelligence, all creativity, and all design, anywhere in the universe, is the direct or indirect product of a cumulative process equivalent to what we here call Darwinian natural

---

30. id., p. 265 (italics added).
31. id., p. 200.

selection. It follows that design comes late in the universe, after a period of Darwinian evolution. Design cannot precede evolution and therefore cannot underlie the universe."[32] Celebrities may believe and voice anything that pops up in their head, but that does not make it science, even if they are professors of the public understanding of science.

"If I were God," said Dawkins in an interview, "I wouldn't do it by evolution! I would do it directly."[33]

The Cultural Animal

Whatever the intentions behind the common effort of scientific materialism to put the human "in his place," the impression that humans are something more or special was too obvious to be completely disregarded. After all, humans write books about science, play the violin, and sell sandwiches. What makes humans different from other primates, biologically speaking, is their bipedalism, which means that they walk upright on two legs, their exceptionally big brain, their ability to speak, and their 'neoteny'. (Neoteny means that the human species retains in adulthood the characteristics of very early childhood compared to the physical development of the other primates.) "On the one hand we are obviously animals comparable with any other. On the other hand we behave quite differently from other animals."[34]

"Up to the human being, the creation works or is realized by communication of the genetic information, something which we have discovered in the middle of the twentieth century," writes Claude Tresmontant. "But from the human being onwards the creation changes its regime, for from the moment when a being appears in the universe that is capable of knowing itself, the new creative information is no longer communicated to its genes. It is communicated to its thought, its intelligence, its mind, its liberty. *This* is the change of regime. The human being is a creation of another order."[35] Owen Gingerich writes: "With the invention of language, humanity crossed the Lamarckian divide from Darwinian evolution, in accordance with which instincts are coded into DNA, to cultural evolution, in which the human brain can begin to store more information than the chromosomes. It is this

---

32. Preface in Susan Blackmore: *The Meme Machine*, p. 9.
33. Russell Stannard: *Science and Wonders*, p. 41.
34. Susan Blackmore: op. cit., p. 1.
35. Claude Tresmontant: *L'histoire de l'univers et le sens de la création*, p. 189.

crossover that makes us distinctly human and uniquely different from the rest of the animal kingdom, the kingdom of which nevertheless we are so much a part."[36]

However much he may have tried, the "divide" remained unlevelled in Richard Dawkins' thinking too. In his dreary landscape crowded with "machines like you and me," lumbering robots and hordes of genes rushing in their blind selfish effort towards the triumph of replication, suddenly appears a vulnerable humanist. "There is no inconsistency in favouring Darwinism as an academic scientist while opposing it as a human being," he writes.[37] One could all the same observe that, if Darwinism and the values of a human being are two different things, Darwinism cannot be *the* complete and universal scientific explanation of life Dawkins wants us to believe it is. (In the previous paragraph Owen Gingerich has already associated the theory with nine lives, Lamarckism, with the cultural side of humanity, and no less an authority than Stephen Gould will do as much.)

"Dawkins' dramatic alternatives – human decency or naked Darwinism – are not posed for effect. He needs the gap between them, because he wants humans to be special. ... The buzz of controversy that surrounds Dawkins in public has obscured the curious likelihood that on questions of evolved psychology, his sympathies may be closer to those of the humanities graduate with whom he shares the op-ed pages than to those of his fellow sociobiologists. ... If not from God or from nature, from where can we derive our sense of how we ought to live? Dawkins is unsure. His ardent moral certainty is not all it appears." (Marek Kohn[38]) Hard as he tried, Dawkins was not able to discard completely the humanistic tradition at the roots of the British academic tradition in which he was educated. Besides, even Bertrand Russell became a public paragon of humanism (and ended with ideas to found a religion).

"We, alone on earth, can rebel against the tyranny of the selfish replicators," writes Dawkins in *The Devil's Chaplain*. The selfish replicators are the selfish genes. (The concept of the "replicator" is one of Dawkins' rhetorical tricks to circumvent the fact that the function of the genes became increasingly more complex and therefore more difficult to use responsibly in his pseudo-Darwinian scheme.) And on the same page we read the astonishing words: "As an academic scientist I am a

---

36. Owen Gingerich: *God's Universe*, p. 107.
37. Richard Dawkins: *A Devil's Chaplain*, p. 13.
38. Alan Grafen and Mark Ridley (eds.): *Richard Dawkins*, pp. 250-1.

passionate Darwinian, believing that natural selection is, if not the only driving force in evolution, certainly the only known force capable of producing the illusion of purpose which so strikes all who contemplate nature. But at the same time as I support Darwinism as a scientist, I am a passionate anti-Darwinian when it comes to politics and how we should conduct our human affairs."[39] The versatile Dawkins is also a study in ambiguity.

Already in the eleventh chapter of *The Selfish Gene*, in which he reveals his theory of the "meme", Dawkins had written: "I am an enthusiastic Darwinian, but I think that Darwinism is too big a theory to be confined to the narrow context of the gene. The gene will enter my thesis as an analogy, nothing more."[40] He wrote this after the glorification of the genes, the legions of blind little gods who create all life on Earth and in the universe. Isidore Nabi will not have been the only one to comment ironically: "In *The Selfish Gene* Dawkins had said that we are 'robot vehicles blindly programmed to preserve the selfish molecules known as genes,' but now he told that we may have to fight against our genetic tendencies. It is really very vexing. Just as I had learned to accept myself as a genetic robot and, indeed, felt relieved that I was not responsible for my moral imperfections, Dr Dawkins tells me that I am not as manipulated as I thought."[41]

After the revolution initiated by Charles Darwin's publication of *The Origin of Species*, Richard Dawkins, like a second Darwin, proudly proclaims his "new revolution" and produces some of his "purple passages" to match the occasion. "Stand tall, Bipedal Ape." This is you. "The shark may outswim you, the cheetah outrun you, the swift outfly you, the capuchin outclimb you, the elephant outpower you, the redwood outlast you. But you have the biggest gift of all: the gift of understanding the ruthlessly cruel process that gave us all existence; the gift of revulsion against the implications; the gift of foresight – something utterly foreign to the blundering short-term ways of natural selection – and the gift of internalizing the entire cosmos."[42]

Even though *The Selfish Gene* was published only one year after Edward Wilson's *Sociobiology*, and both works shared the same inspiration and apparently propagated the same doctrine, the dominant stresses and their whole outlook were quite different. "It is hard to regard

---

39. Richard Dawkins: op. cit., p. 13.
40. id.: *The Selfish Gene*, p. 191.
41. Ullica Segerstråle: op. cit., p. 184.
42. Richard Dawkins: op. cit., p. 15.

Wilson's sociobiology as being similar to Dawkins','" writes Ullica Segerstråle. "The core of sociobiology (as we now know it) was articulated
in *The Selfish Gene* rather than in Wilson's *Sociobiology*. Wilson's book
was called 'the new synthesis', but for practicing sociobiologists the
ideas presented in Dawkins' book became the synthesis-in-use. And
the concept that helped delineate and solidify the new sociobiological
paradigm was the gene's eye view."[43] Wilson, with his all-pervading
scientism, could not accept the "divide" between the physical realm of
the genes and the cultural realm of the 'memes', each with analogous
but parallel mechanisms. "Wilson could not really 'afford' to leave [as
Dawkins did] the cultural realm as a separate one, sitting on top of
the genetic one, even if he entertained this as a theoretical possibility:
materialism had to be guaranteed."[44]

Memes – The Gremlins of the Mind

The "memes" are another road-side attraction in the Dawkinsian
wonderland. The origin of the idea has been sketched in the previous section and is a certainly a valid topic of reflection and eventual
research, even if it is grotesque that scientism has to make an effort to
recognize that the human being belongs to a category of its own. "Are
there any good reasons for supposing our own species to be unique? I
believe the answer is yes," declares Richard Dawkins.[45] The problem
with him and his ideological brothers-in-arms, however, is their usual
"full-throated" affirmation of what is most often no more than a vague
intuition, while on the next page or in their next book one will find a
negation or the contrary of their own 'scientific' novelties.

Once more we find the seed of a Dawkinsian venture in Jacques
Monod. "Though the abstract realm transcends the biosphere even
more than the biosphere transcends the lifeless universe, the ideas have
retained certain properties of the organisms." (Here Monod bases his
thought clearly on the three traditional realms of "lifeless" matter, life,
and mind, which he calls "the abstract realm.") "Just like the organisms,
the ideas tend to perpetuate and to multiply their structure; like the organisms they are able to fuse, to recombine, to segregate their contents;

---

43. Alan Grafen and Mark Ridley (eds.): op. cit., pp. 85, 91.
44. Ullica Segerstråle: op. cit., p. 40 (italics added).
45. Richard Dawkins: op. cit., p. 189.

like the organisms they evolve, and in this evolution [natural] selection doubtlessly plays an important role." Monod compares the realm of the mind to the realm of the living organisms, and sees in both an identical or at least an analogous way of evolving. He concludes: "I would not be so bold as to propose a theory of [natural] selection of the ideas, but one could at least try to define some of the principal factors which play a role in it."[46]

Where Monod feared to tread, Dawkins rushed in at the end of the book that would make him famous, *The Selfish Gene*. The last chapter of the original edition was titled: "Memes: the new replicators." It may be recalled that "replicator" is the abstract term used by Dawkins to evade the increasingly difficult definition of the gene and its functions. "A replicator is a piece of coded information that makes exact copies of itself," explains Dawkins. "The archetypal replicator is a gene, a stretch of DNA that is duplicated nearly always with extreme accuracy, through an indefinite number of generations."[47] It may be noted that in the first sentence the piece of coded information copies, or replicates, itself, while in the following sentence it is "duplicated," which is the correct statement. According to Denis Noble one should "avoid saying that genes do anything at all; it is more that genes are used." We will have to come back to this important and generally misunderstood point.

"I think that a new kind of replicator has recently emerged on this planet. It is staring us in the face. It is still in its infancy, still drifting clumsily about in its primeval soup, but already it is achieving evolutionary change at a rate that leaves the old gene panting far behind. The new soup is the soup of human culture." Thus writes Dawkins in his baptismal oration of the memes. The soup metaphor is used in analogy with the origin of life, the first evolution; the "new soup" is the mother-broth of the second evolution, that of the mind, when the siblings with the big brains appeared among the primates. "Recently" means here from one to two million years ago, a short time-span in evolution.

"We need a name for the new replicator," continues Dawkins, "a noun that conveys the idea of a unit of cultural transmission, or a unit of *imitation*. 'Mimeme' comes from a suitable Greek root, but I want a monosyllable that sounds a bit like 'gene'. I hope my classicist friends will forgive me if I abbreviate mimeme to *meme*. ... It should be pronounced to rhyme with 'cream'. Examples of memes are tunes, ideas,

---

46. Jacques Monod: op. cit., pp. 208-9.
47. Richard Dawkins: *The God Delusion*, p. 191.

catch-phrases, clothes fashions, ways of making pots or building arches. Just as genes propagate themselves in the gene pool by leaping from body to body via sperm or eggs, so memes propagate themselves in the meme pool by leaping from brain to brain via a process which, in the broad sense, can be called imitation. ... When you plant a fertile meme in my mind you literally parasitize my brain, turning it into a vehicle for the meme's propagation in just the way that a virus may parasitize the genetic mechanism of a host cell."[48]

Dawkins' idea of the meme has already parasitized a considerable number of brains, contaminating them with the meme virus and the fever to write articles and books about it, thus passing it on. It has also succeeded to invade the Oxford Dictionaries, where we find its definition, "a cultural or behavioural element passed on by imitation or other non-genetic means."[49] Other definitions are e.g. "the fundamental unit of cultural transmission ... a mental item that is borrowed from one person and passed on to another ... a replicator that jumps from one brain to another ... whatever it is that is passed on by imitation ... a unit of cultural inheritance... a unit of information, with a definite structure, residing in the brain ..." The last two definitions are by Dawkins himself.

"Both genes and memes are replicators and must obey the general principles of evolutionary theory and in that sense are the same. Beyond that they may be, and indeed are, very different," writes Susan Blackmore. "They are related only by analogy."[50] The moon and the face of the beloved, termite nests and skyscrapers, vaulting poles and redwoods, children and angels – all are easy to relate by analogy, but less easy to relate by science. The analogy at the core of the meme idea is the one quoted above: "Just as genes propagate themselves in the gene pool by leaping from body to body, via sperm and eggs, so memes propagate themselves in the meme pool by leaping from brain to brain." But genes are very material structures of DNA molecules, memes are invisible because immaterial . Genes are transmitted by intercourse. How are the jumping memes caught? A gene pool is the conglomerate of the genomes of a group of organisms. What is a meme pool? A meme is said to have a definite structure. Which one? The questions are many and remain unanswered, or unanswerable.

Susan Blackmore, who wrote a book "to lay the foundations for a

48. Richard Dawkins: op. cit., p. 192.
49. *The Concise Oxford Dictionary*, tenth edition.
50. Susan Blackmore: op. cit., p. 17.

science of memetics" (Richard Dawkins wrote the preface), recognizes that there are problems with memes. She mentions three. "1. We cannot specify the unit of a meme. – I have heard people dismiss the whole idea of memetics on the grounds that 'you can't even say what the unit of a meme is'. Well that is true, I cannot. And I do not think it is necessary. 2. We do not know the mechanism for copying and storing them. – No we do not. ... We may get a long way with the general principles of memetic selection without understanding the brain mechanisms it relies on. We can also make some educated guesses about these mechanisms based on what little we know. 3. Memetic evolution is 'Lamarckian'. – Lamarck believed in all sorts of things that have been rejected. ... The whole idea of Lamarckian inheritance is irrelevant. ... My conclusion about Lamarck is that the question 'Is cultural evolution Lamarckian?' is best not asked." (The author warns us: "To start to think memetically we have to make a giant flip in our minds". And she takes us into her confidence: "A book may seem very much like a [meme] vehicle, but my pumpkin soup does not. I am not at all sure where to draw the lines here."[51])

"The haven all memes depend on reaching is the human mind, but a human mind is itself an artifact created when memes restructure a human brain in order to make it a better habitat for memes," asserts Daniel Dennett. Dennett is a nowadays widely read philosopher whom Gould has called "Dawkins' lapdog." It is true that Dawkins has spent affectionate praise on him. It stands to reason that scientists of the Dawkins trend like a philosopher who proffers statements like the one just quoted, or: "The physics and chemistry of life are now understood in dazzling detail,"[52] which is at least slightly exaggerated.

The same Dennett writes: "The prospects for elaborating a rigorous science of memetics are doubtful, but the concept provides a valuable perspective from which to investigate the complex relationship between cultural and genetic heritage."[53] And: "At this time the contributions of the concept of the meme are still largely conceptual – or philosophical."[54] This suggestion that the philosophy of the meme is an activity practiced in the clouds would have had Aristophanes' consent.

---

51. Susan Blackmore: op. cit., pp. 53 ff., 7, 65.
52. Daniel Dennett: *Darwin's Dangerous Idea*, p. 150.
53. id., p. 369.
54. Alan Grafen and Mark Ridley (eds.): op. cit., p. 114.

It is rather puzzling to learn in Dawkins' preface to Blackmore that Dennett has become "the philosophical mentor of all meme theorists."[55]

It should however be realized that the intuition at the origin of the meme theory is a valid one. It rests on the indelible awareness that the human being is not just an animal but a thinking animal, whose mind puts it in a category apart. This obstinate fact cannot be accepted by scientific materialism to which mind is nothing but an aspect, or effect, or 'epiphenomenon', of matter. Dawkins has tried to integrate the mind and its functions into the materialistic doctrine. But materialistic science, despite it being a mental exercise, does not possess the ideological instrumentation to do so, with the result that Dawkins meme theory is for the most part tentative and high-stilted verbiage.

The critics of memetics point this out, for instance Alister McGrath, a theologian with a PhD in molecular biophysics. He writes: "1. There is no reason to suppose that cultural evolution is Darwinian, or indeed that evolutionary biology has any particular value in accounting for the development of ideas. 2. There is no direct evidence for the existence of 'memes'. 3. The case for the existence of the 'meme' rests on the questionable assumption of a direct analogy with the gene, which proves incapable of bearing the theoretical weight that is placed upon it. 4. There is no necessary reason to propose the existence of a 'meme' as an explanatory construct. The observational data can be accounted for perfectly well by other models and mechanisms."[56]

"No significant body of empirical research has grown up around the meme concept," writes Robert Aunger, "nor has memetics made empirically testable propositions or generated much in the way of novel experimental or observational data. In fact, the memetic literature remains devoted almost exclusively to theoretical antagonisms, internecine battles, and scholastic elucidations of prior writings on memes. This is typically the sign of a science in search of a subject matter. ... I predict that memetics is unlikely ever to become an empirical science, because when we define memes in a manner precise enough to start making testable predictions, we find that we have largely defined them out of existence."[57]

---

55. Susan Blackmore: op. cit., p. xvi.
56. Alister McGrath: *Dawkins' God – Genes, Memes, and the Meaning of Life.* p. 121.
57. Alan Grafen and Mark Ridley (eds.): op. cit., pp. 178, 186.

A Devil's Chaplain

The response to Dawkins' selfish gene theory has been sensational. The reason of its ready acceptance is quite obvious: selfishness, or ego, is evident as the root of most behaviour in the realm of life. His 'memes' have been received more cautiously, although this thesis too has already spawned a considerable literature. But he is most famous among the general public for his direct and sustained attack on religion and the concept of God. The subtitle of *The Blind Watchmaker* (1986) is "Why the evidence of evolution reveals a universe without design," with natural theology as its main target. The collection of his essays in *A Devil's Chaplain* (2003) takes up the cause against a creator or a creative intelligence behind evolution. In *The God Delusion* (2006) he bundles his arguments in a direct attack on a Supreme Being and any form of religion.

Richard Dawkins is a famous man, even more so than the late Carl Sagan, the glamour boy of exobiology (extraterrestrial life). Dawkins is called "the best popularizer of Darwinism around" and "the number one public intellectual in England." Microsoft billionaire Charles Simonyi has funded a special chair for him in the Public Understanding of Science at Oxford. He has been made a fellow of the Royal Society, although "it is noteworthy that Dawkins' election to the Royal Society was for his contribution to the public understanding of science, not for his contribution to science itself." (Alan Grafen) No TV programme, no conference, no seminar, no panel of importance about the relation of science and religion, or science versus religion, without the presence of Professor Dawkins. All this, added to his articles, books and remunerative awards, makes him one of the most distinct voices in the choir of public science. "It's all sex or money or genes. A simple and dramatic theory that explains everything makes good press, good radio, good TV, and best-selling books. Anyone with academic authority, a halfway decent writing style, and a simple and powerful idea has easy entry to the public consciousness," ponders Richard Lewontin.[58]

Dawkins is also called "prophet of science," "chief gladiator against religion" and "the world's most high-profile atheist polemicist." Alister McGrath notes that "one of the most melancholy aspects of *The God Delusion* is how its author appears to have made the transition from a scientist with a passionate concern for truth to a crude anti-religious

---

58. Richard Lewontin: *The Doctrine of DNA*, p. vii.

propagandist who shows a disregard for evidence."[59] As this are the words of a Catholic theologian, bias may be surmised. In this case it should not be. "I have never heard such hard-line, aggressive promotion of atheism under the guise of science as I have heard from the Darwinists," writes Denyse O'Leary. – Remember: Dawkins is considered "Darwinist-in-chief." – "It is, at best, amusing to hear Darwinists charge that the creationists have an underlying religious agenda, when the Darwinists' own anti-religious agenda is pretty obvious."[60]

"... To be an atheist is a realistic aspiration, and a brave and splendid one. You can be an atheist who is happy, balanced, moral, and intellectually fulfilled," assures Dawkins.[61] "For Dawkins, Darwin was a revelation. Dawkins was a zoologist, and ethologist, and then suddenly Darwin got to him, and he thought, My God, this is the truth, and everybody should know this truth! He became something like a preacher."[62] In *The Devil's Chaplain*, Dawkins did indeed declare: "Darwin's achievement, like Einstein's, is universal and timeless," and we have already expressed the opinion that behind his scientific formation the character can be sensed of an enthusiast who might turn into a zealot. "For Dawkins," writes McGrath, "Darwin's theory of evolution is more than a scientific theory. It is a worldview, a total account of reality. Darwinism is a 'universal and timeless' principle, capable of being applied throughout the universe."[63]

"Personal incredulity screams from the depths of my prescientific brain centres," Dawkins exclaims[64] – incredulity in religious, supernatural and paranormal matters, that is. According to McGrath, who has written *The Dawkins Delusion* to countermand *The God Delusion*, four interconnected grounds of hostility may be found throughout his writings: 1. a Darwinian worldview makes belief in God unnecessary; 2. truth is grounded in explicit [scientific] proof; 3. religion offers an impoverished and attenuated vision of the world; 4. religion leads to evil.

In an interview, Dawkins happily declares himself to be a zealot for his conviction. "I like to think of myself as a creative force in the field [of ideology] ... We do something creative, we change people's minds. My zealotry comes from a deep concern for the truth. I'm extremely

---

59. Alister McGrath: *The Dawkins Delusion*, p. 27.
60. Denyse O'Leary: *By Design or by Chance?* p. 239.
61. Richard Dawkins: op. cit., p. 3.
62. John Brockman: *The Third Culture*, p. 88.
63. Alister McGrath: op. cit., p. 43.
64. Richard Dawkins: *The God Delusion*, p. 129.

hostile towards any sort of obscurantism, pretension."[65] In the opening pages of *The God Delusion* we read: "If this book works as I intend, religious readers who open it will be atheists when they put it down."

"Some of the most spirited and vocal defenders of evolutionary theory, such as Richard Dawkins, use their stature as scientific spokes-men as a bully pulpit for atheism," regrets Owen Gingerich, who is a religious scientist.[66] Other such spokesmen are many. Christopher Hitchens, for instance, addresses the readers somewhere in the middle of his Dawkinsian book as follows: "Dear reader, if you have come this far and found your faith undermined – as I hope ..."[67] Another kin-dred author is Daniel Dennett, who wrote *Breaking the Spell – Religion as a Natural Phenomenon* "to get his readers to think and not just to feel." His central policy recommendation is "that we gently, firmly educate the people of the world, so that they can make truly informed choices about their lives."[68] At last the world re-educators have come, but that "gently, firmly" may make some feel uneasy. "Edward Wilson and other scientists have promoted [the sociobiological] model of human nature in popular books and magazines with missionary fervour, aiming to convert the unenlightened. ... Their claims, their language and their style have striking religious overtones. Some sociobiologists like Rich-ard Dawkins pride themselves on being materialist, reductionist and overtly anti-religious. But they offer theories proclaiming the evolution-ary basis of human behaviour as explanations for virtually everything and as the basis for the unification of knowledge. Scientists promoting genetic explanations use a language replete with religious metaphors and concepts such as immortality and essentialism – indeed, the gene appears as a kind of sacred 'soul'. And as missionaries bringing truth to the unenlightened, they claim their theories are guides in moral action and policy agendas. They are part of a current cultural move to blur the boundaries between science and religion," writes Dorothy Nelkin.[69]

In the same essay she continues: "The language used by geneticists to describe the genes is permeated with biblical imagery. Geneticists call the genome 'the Bible', 'the Book of Man' and 'the Holy Grail'. They convey an image of this molecular structure as more than a powerful biological entity: it is also a mystical force that defines the natural and

---

65. John Brockman: op. cit., p. 85.
66. Owen Gingerich: *God's Universe*, p. 74.
67. Christopher Hitchens: *God is not Great*, p. 182.
68. Daniel Dennett: *Breaking the Spell*, p. 339.
69. Hillary and Steven Rose (eds.): *Alas Poor Darwin*, p. 15.

moral order. And they project an idea of genetic essentialism, suggesting that by deciphering and decoding the molecular text they will be able to reconstruct the essence of human beings, unlock the key of human nature. ... Such images fuel popular narratives of genetic essentialism – a picture of the gene as the essence of the person, the locus of good and evil, the key to 'the secret of life'."[70]

But sometimes the muzzle comes off. "I am attacking God, all gods, anything and everything supernatural, wherever and whenever they have been or will be invented," warns Dawkins. "The God of the Old Testament is arguably the most unpleasant character in all fiction: jealous and proud of it; a petty, unjust, unforgiving control-freak; a vindictive, bloodthirsty ethnic cleanser; a misogynistic, homophobic, racist, infanticidal, genocidal, filicidal, pestilential, megalomaniacal, sadomasochistic, capriciously malevolent bully."[71] (Dawkins also spoke the often quoted words: "If you meet somebody who claims not to believe in evolution, that person is ignorant, stupid or insane – or wicked, but I'd rather not consider that." More recently he added: "I don't withdraw a word of my initial statement. But I do now think it may have been incomplete. There is perhaps a fifth category, which may belong under 'insane' but which can be more sympathetically characterized by a word like tormented, bullied, or brainwashed."[72])

Yet Dawkins is not the only scientist to tilt his lance against religion and God. Physicist and Nobel Prize winner Steven Weinberg, for instance, has said: "Religion is an insult to human dignity. With or without it, you'd have good people doing good things and evil people doing evil things. But for good people to do evil things, it takes religion."[73] It has been referred to in the previous chapters that the gradual downslide of religion among the brightest minds is a Western phenomenon, countering the centuries of dogmatic tyranny by the Catholic Church, followed in this by the principal Protestant churches. This is the reason why the God under attack is without exception the Judeo-Christian God – most of the anti-theists have no idea of any other. A concordant reason of their aggressive stance is the twentieth century American brand of Christian fundamentalism and creationism, causing the trials about which world interpretation should be taught in the schools.

---

70. id., p. 18.
71. Richard Dawkins: op. cit., pp. 36, 31.
72. William Dembski (ed.): *Uncommon Dissent*, xvii.
73. Richard Dawkins: op. cit., p. 249.

10.

The Darwin Wars

Another curious aspect of the theory of evolution is that everybody thinks he understands it.

JACQUES MONOD

Stephen Gould and Punctuated Equilibrium

Stephen Jay Gould (1941-2002) and Richard Dawkins had much in common. Gould "filled the same role for North America as Dawkins does for Britain."[1] Both were evolutionary biologists, deeply involved as students, researchers and teachers in the academic world; they were talented, prolific and successful writers; and they were critical of Charles Darwin even when professing their appreciation of him. Gould, who declared his "love for Darwin and the power of his genius," has been called "America's leading evolutionary biologist" and "North America's Darwinist-in-chief." He has written a great number of books which are still widely read, and therefore continue to be influential with the experts as well as among the informed public. Some of these books are selections from the essays he regularly published in the science magazine *Natural History. Ever since Darwin, The Panda's Thumb, Wonderful Life, Hen's Teeth and Horse's Toes* and *Rock of Ages* are a few of the titles by this cultured writer.

Gould is best known for his hypothesis of "punctuated equilibrium." This at first sight mystifying term is not a symptom of the labile sense of balance typical of people under the influence. The theory it covers is a direct criticism and contradiction of a pillar of Darwin's theory of evolution: gradualism. "[Niles] Eldredge and I coined the term punctuated equilibrium in a paper first presented in 1971," wrote Gould. "Our original article on punctuated equilibrium (Eldredge and Gould, 1972)

---

1. Denyse O'Leary: *By Design or by Chance?* p. 108.

161

emerged as a result. ... I did coin the term punctuated equilibrium, but the basic structure of the theory belongs to Eldredge."[2]

In the evolution of species "discontinuities are overwhelmingly frequent ... The discontinuities are even more striking in the fossil record. New species appear in the fossil record suddenly, not connected with their ancestors by a series of intermediates. Indeed there are rather few cases of continuous series of gradually evolving species," states Ernst Mayr, one of the stalwarts of neo-Darwinism.[3] This does not agree with Darwin's original theory which posited that "as natural selection acts solely by accumulating slight, successive, favourable variations, it can produce no great or sudden modifications; it can act only by short and slow steps. ... If it could be demonstrated that any complex organ existed which could not have been formed by numerous, successive, slight modifications, my theory would absolutely break down. ... Natural selection acts only by taking advantage of slight successive variations; she can never take a great and sudden leap, but must advance by short and sure, though slow steps."[4]

Darwin's conviction could not be expressed more clearly. (Equally clear is the fact that, if a theorist holds the opposite view, he cannot claim to be a Darwinian.) However, the gaps in the fossil record are undeniable. They were already a constant headache for Charles Darwin himself. His theory of variations and natural selection demanded that "an interminable number of intermediate forms must have existed ... infinitely many fine gradations between past and present species ... The abrupt manner in which whole groups of species suddenly appear in certain [geological] formations, has been urged by several paleontologists as a fatal objection to the belief in the transmutation of species. If numerous species, belonging to the same genera or families, have really started into life at once, the fact would be fatal to the theory of evolution through natural selection."[5] Darwin knew how essential this point was for his theory, for in the *Origin* he returned to it again and again, trying to justify the gaps in the fossil record by the lack of relevant paleontological discoveries or the destruction of the fossils through natural causes.

Stephen Gould was an expert in paleontology, who taught it and worked at the American Museum of Natural History in New York. He

2. Denyse O'Leary: op. cit., p. 51.
3. Ersnt Mayr: *What Evolution Is,* p. 208.
4. Charles Darwin: *The Origin of Species*, pp. 415, 171, 181.
5. id., pp. 429, 305.

found that, 150 years after the publication of the *Origin*, the insufficiency of the fossil record, "the best kept trade secret" in biology, could no longer be adduced as an argument to explain the gaps. He lashed out at the theory of gradualism, which originally was a geological theory and against which Thomas Huxley and Alfred Wallace had cautioned Darwin from the start. Gould wrote that the "assumptions of gradualism had stymied and constrained our comprehension of the earth's much richer history." He found himself "forced *to question the necessary basis for Darwin's key assumption* that observable, small-scale processes of microevolution could, by extension through the immensity of geological time, explain all patterns in the history of life."[6]

In the fossil record Darwin's "infinitely many fine gradations between past and present species" were not there. "It is a feature of the known fossil record that most taxa [branchings-off] appear suddenly. They are not, as a rule, led up to by a sequence of almost imperceptibly changing forerunners such as Darwin believed should be usual in evolution. A great many sequences of two or a few temporally intergrading species are known, but even at this level most species appear without intermediate ancestors, and really, perfectly complete sequences of numerous species are exceedingly rare." In other words: transitions from one species to another are very rarely embodied in intermediary variations: species appear nearly always suddenly. Moreover: "The geological record is extremely spotty ... extremely imperfect ... The fossil record may, after all, be 99 percent imperfect."[7]

"Eldredge and Gould looked at the fossil record. What happens? It's absolutely startling! You don't get one species turning into another. A species emerges, it lasts for several million years, and it disappears. Five hundred million years for some of these species, a few million for others. Species emerge suddenly, not slowly. Punctuated equilibrium keeps these problems of emergence in focus." (Brian Goodwin[8]) Punctuated equilibrium "merely honoured the firmest and oldest of all paleontological observations."[9] Looking objectively at the available fossils, the gaps in the hierarchical classification had always been there, but they had been overlooked or denied by prejudice inspired by the Darwinian dogma.

"Punctuated equilibrium makes the strong claim that, in most cases,

---

6. Stephen Jay Gould: *Punctuated Equilibrium.*, pp. 9-10 (italics added).
7. id., pp. 26, 29, 31.
8. John Brockman: *The Third Culture*, p. 101.
9. Stephen Jay Gould: op. cit., p. 38.

effectively no [variational or mutational] change accumulates at all. A species at its last appearance before extinction does not differ systematically from the anatomy of its initial entry into the fossil record, usually several million years before."[10] Should it differ? According to the Darwinian doctrine it should, for the simple reason that natural selection is omnipresent and continuous. As Darwin wrote: "Natural selection is daily and hourly scrutinizing, throughout the world, every variation, even the slightest."[11] The fact that species remain the same during millions of years, from their appearance to their extinction, Gould and Eldredge have called "stasis", a Greek word which means something like 'steadfastness'.

A famous example of stasis is the coelacanth, a species of fish which originated 400 million years ago and which is still alive and swimming. The coelacanth predates the dinosaurs by more than 200 million years and was thought to have become extinct with them, as its most recent fossils dated from 65 million years ago. A living specimen of the coelacanth was discovered in 1938, in the waters close to the Comoro Islands. Afterwards more living specimens were found, and in 1987 coelacanths were filmed from a submersible in their natural habitat. The discovery of the coelacanth was considered "the biological find of the century." One reason for this hyperbole was that at the time many biologists thought the coelacanth to be the crucial link between fish and amphibians. Another reason was that the amazing coelacanth seems hardly to have changed in all those millions of years, a permanence contradicting the continuous effect of Darwinian natural selection. Of this stasis, the coelacanth is now known to be only one example among many.

Little by little the theory of evolution has made room for "macroevolution" side by side with "microevolution." The latter, evolution in small (micro) steps, is the only one Darwinism and neo-Darwinism accept, for its mechanism is the selection of variations, later on found to be genetic mutations. Macroevolution, on the contrary, is evolution by big (macro) steps, causing the appearance of new species suddenly. Remember the words of Ernst Mayr: "New species appear in the fossil record suddenly, not connected with their ancestors by a series of intermediates." According to Gould's thesis "new species originate in a geological 'moment'." The scale of geological periods covers many

---

10. id., p. 41.
11. Ernst Mayr: op. cit., p. 155.

millions of years. The 'moment' in which a new species takes shape may last from thousands to several million years. "Species appear to change little, or not at all, during their lifetimes ... Each speciation event occurs quite rapidly in geological terms, so rapidly that it has sometimes been called 'quantum speciation', on analogy with the 'quantum jumps' that occur in atoms and molecules." (Burton Guttman[12])

The meaning of "punctuated equilibrium" should now be clear. Species do not change markedly during their existence, their shape remains static; and instead of the theoretical evolutionary change required by natural selection, there is stasis, which is a form of equilibrium. On the other hand, the fossil record shows that species appear suddenly, at a geological moment in time; the existing equilibrium is punctuated by short-term formations of new species. This, again, is in contradiction with the classical Darwinian doctrine, which requires that they should evolve gradually, in small steps, causing small variation after small variation which finally lead, "over time," to a fully formed new species. In the Darwinian view time is the magician who produces and explains all biological wonders.

The reason why Charles Darwin stuck doggedly to his tenet of gradualism was that he could not imagine another way for organisms to change. If an evolutionary 'mechanism' existed to cause "modification" – and if the creation of new species was not an intervention by an non-material, supernatural Creator or Intelligence – it could, Darwin thought, only work in very small steps, gradually. (How this 'mechanism' actually functioned was not known, and its existence was therefore nothing but guesswork. "Very small" meant mechanically possible and therefore supposedly feasible by nature; not "very small" seemed to require a miracle.) That speciation has happened in big sudden steps, by macroevolution, seems to be the only correct interpretation of the fossil record. But how could such big steps, requiring astronomical numbers of simultaneous genetic and physical modifications, come about at once, in a geological moment, even if this moment lasted for thousands or a few millions of years? However strongly Stephen Gould affirmed that he was "an old-fashioned materialist," his thesis smelled of creationism and was consequently violently attacked – though at present, it seems to have gained the upper hand.

---

12. Burton Guttman: *Evolution*, p. 57.

Mass Extinctions and the Dinosaurs

The thesis of Gould and Eldredge was mainly based on the data of paleontology. That it drew the attention away from genetics and molecular biology in general, was another reason of the adverse reactions it caused. Although the fossil collections had been considerably augmented since Darwin's day, the record as a whole remained "extremely spotty." But even in its spottiness it supported some striking facts. One was the phenomenon of "the Cambrian explosion", called "the most remarkable episode in animal evolution."

The geological period called the Cambrian lasted from 542 million to 488 million years ago. To the astonishment of the paleontologists, practically all "body plans" originated then. A body plan in biology is the basic way in which organisms have been built, e.g. with or without a backbone, or bodies made like those of spiders, or wasps, or worms, or mainly consisting of 'jelly'.

Another striking fact was the scale of the extinctions of organisms that have taken place on our planet. Beside frequent minor ones, five major extinctions have caused the disappearance of no less than 99.9 percent of all living beings which have ever existed. Those five catastrophes, occurring each time towards the end of the respective geological periods, happened in the Ordovician (-440 million years, 85 percent of all life forms), the Devonian (-365 million years), the Permian (-251 million years, the most devastating, no less than 96 percent of all life forms perished), the Triassic (-205 million years, 76 percent), and the Cretaceous (-65 million, 75 to 80 percent, including the dinosaurs).

As to the dinosaurs, the common notion is that they were destroyed when an asteroid hit the Earth in the Gulf of Mexico. The evidence which made the explanation supposedly irrefutable was provided by Walter Alvarez. "No one now seriously doubts that there was a meteor impact at the Cretaceous/Tertiary boundary. But there is still a lot of debate on its significance. After all, if that is all that happened, why did crocodiles, turtles and even frogs go through relatively unscathed?" asks Kim Sterelny.[13] "It was a physicist [Alvarez] who pushed most dogmatically the view that the dinosaurs were killed off in a discrete event by the simple cause of the collision with Earth of an asteroid ... To paleontologists, however, that seems absurdly oversimplified," writes

---

13. Kim Sterelny: *Dawkins vs. Gould*, p. 109.

Henry Bauer.[14] And in Gribbin we find: "The dinosaurs dwindled over a span of at least 10 million years, from 30 genera to 13 found in the fossil beds of Montana and southern Alberta. And the 'terminal event' was not all that terminal. ... The dinosaurs did not die out overnight ... There was a gradual decline in the number of dinosaur (and other) species over millions of years."[15] Francis Hitching mentions no less than six possible events which may have caused the dinosaurs' disappearance. Which means that in biology another legend was added to the substantial list we have already encountered.

Of Human Arrogance

"I am an old-fashioned materialist," said Stephen Gould. "I think the mind arises from the complexities of neural organization, which we don't really understand very well."[16] The accumulating data of the fossil record may have contributed to his increasingly dark view about the meaning of the whole shebang, of human life no less than of a cuckoo's egg and the countless beings that have peopled and are peopling our planet. Gould's view of the evolution and the place of humanity in it is the logical one when all values have evaporated. Still, it is noteworthy that the devaluation of the human existence, based on "the Copernican principle" or "the principle of mediocrity," runs in the biological sciences parallel to its re-evaluation in physics, where the "anthropic principle" is presently one of the hot topics. The former wants to counter the age-long human arrogance of occupying a special spot in the evolution of life on Earth, while its existence is scientistically speaking a matter of pure chance; the latter, after having arduously tried to do as much, has recently been awestruck by the magnitude of pure chance in the history of the universe, of which the existence of our species is the result.

"We are here," Gould is quoted as saying, "because an odd group of fishes had a peculiar fin anatomy that could transform into legs for terrestrial creatures; because the earth never froze entirely during an ice age; because a small and tenuous species, arising in Africa a quarter of a million years ago, has managed, so far, to survive by hook and by

---

14. Henry Bauer: *Scientific Literacy and the Myth of the Scientific Method*, p. 27.
15. Mary and John Gribbin: *Being Human*, pp. 96, 70, 76.
16. John Horgan: *The End of Science*, p. 125.

crook. We may yearn for a 'higher' answer – but none exists."[17] "The earliest known vertebrate located in the Burgess shale [in the middle Cambrian] is a two-inch and rather elegant creature named, after an adjoining mountain and also for its sinuous beauty, *Pikaia gracilens*. It was originally and wrongly classified as a worm (one must never forget how recent much of our knowledge really is), but in its segments, muscularity, and dorsal-rod flexibility, it is a necessary ancestor of *Homo sapiens* ... Millions of other life forms perished before the Cambrian period was over, but this little prototype survived. To quote Gould: '... If *Pikaia* does not survive, we are wiped out of future history – all of us, from shark to robin to orangutan.'"[18]

If there are no values, it is not possible to evaluate "progress" in the unfolding of life. "Gould was adamantly opposed to progress, speaking of it as 'a noxious, culturally embedded, untestable, nonoperational, untractable idea that must be replaced if we wish to understand the patterns of history'. It is a delusion engendered by our refusal to accept our insignificance when faced with the immensity of time."[19] That evolution is not progressive became one of the main themes of his work, as zealously preached as the gospel according to Richard Dawkins. In this he was once more an apostate from Darwinism, for the Master had written towards the end of the *Origin*: "We may look with some confidence to a secure future of great length. And as natural selection works solely by and for the good of each being, all corporeal and mental endowments will tend to progress towards perfection." True to character, however, the Master had also uttered his doubts: "Natural selection, or the survival of the fittest, does not necessarily include progressive development ..."[20]

That a biologist, and one with authority, does not necessarily have to turn into a zealot for the scientist faith, is shown e.g. by Theodosius Dobzhansky, one of the architects of neo-Darwinism and a Russian Orthodox Christian. He wrote: "Seen in retrospect, evolution as a whole doubtless had a general direction, from simple to complex, from dependence on to a relative independence of the environment, to greater and greater autonomy of individuals, greater and greater development of sense organs and nervous systems conveying and processing information about the state of the organism's surroundings, and finally greater

---

17. http://homepage.eircom.net/odyssey/quotes/life/science/evolution.html
18. Christopher Hitchens: *God is not Great*, p. 109.
19. id., p. 145.
20. Charles Darwin: op. cit., pp. 450, 124.

and greater consciousness. You can call this direction progress or by some other name."[21]

The Darwin Wars

"There is no more contentious, querulous bunch of professionals on earth than evolutionary biologists," writes Arthur Shapiro, who is an evolutionary biologist himself.[22] It is safe to say that in the history of science no new discovery, concept or theory has seen the light of day without being attacked, at times viciously. What has come to be called the Darwin wars "are not between believers and disbelievers in evolution, or in Darwinism. They are about the scope and proper limits of Darwinian explanation."[23] Nonetheless, if the controversies have been so nasty that "the enmities made in that struggle persist to this day," their stakes must have had deeper than theoretical roots. Andrew Brown is of the same opinion, for the subtitle of his book *The Darwin Wars* is "the scientific battle for the soul of man."

The first confrontation in the Darwin wars was triggered off by the publication of Edward Wilson's *Sociobiology*. We know the extremist, totalitarian ambition of Wilson's scientism. He saw in science, the ultimate explanation of existence, a substitute for religion, and in evolution the foundation of all animal behaviour, including that of the human animal. Science, for sure, is at its core a search for Truth. However, the mistake it has made and continues making time and again is that a partial discovery of that Truth is held to be its totality, although even a casual glimpse of the history of science should teach anybody that "the path of science is lined with the corpses of dead theories."

Wilson's sociobiology was a form of social Darwinism, which had also engendered eugenics. After the global wave of eugenics in the first half of the 20th century, and after the horror of its excesses by the Nazis had become known, social Darwinism was hastily abandoned by the biologists who were its proponents. But Wilson's stance in 1975 was a clear reintroduction of the same evolutionary and biological principles. The reaction of the leftist fringe in American science, lead by the very able geneticist Richard Lewontin and the zoologist Stephen Gould, was

---

21. http://homepage.eircom.net/odyssey/quotes/life/science/evolution.html
22. William Dembski (ed.): *Uncommon Dissent*, p. 292.
23. Andrew Brown: *The Darwin Wars*, p. 18.

immediate, fierce, and waged in the media. Gould "hated sociobiology" for its deterministic view of the human species, contradicting his humanitarian socialism for which progress, if not evolutionary, was politically possible. One of the highlights of this battle of the intellectuals was "the ice water incident" when in 1978 leftist hecklers emptied a pitcher of water on Edward Wilson's head.

Now that we know of Dawkins' launch of the meme theory, intended as a reaction against absolute genetic determinism, and how much Gould and Dawkins, American and British self-professed descendants of Darwin, had in common, it seems difficult to comprehend that they became the leaders of the opposing camps in the second phase of the Darwin Wars. But we have already seen that Dawkins' reasoning rested mainly on theories about the genes, their "lineages" and their alleged action in the extended phenotypes, while Gould's vantage point was primarily the study of the fossil record. (The various disciplines assembled under the umbrella of 'biology' are quite different. Anatomy is a matter of dissection of organisms, taxonomy of their classification, ethology of their behaviour, genetics of their heredity, paleontology of their history, etc.) Moreover, the intellectual backgrounds of Dawkins and Gould varied significantly, for the revolutionary 60s, in which they reached their maturity, had been mainly an American phenomenon.

In the Dawkins camp there were Edward Wilson, William Hamilton, John Maynard Smith, Daniel Dennett, Steven Pinker, Peter Atkins ... – names known from the covers of popular science literature. "Most are passionately anti-religious, or at least passionately opposed to modern Protestantism, which, like its adherents, they take to be the only true religion." On Gould's side there were Richard Lewontin, Steven Rose, and many of the younger scientists who had grown up during the turbulent years of the Vietnam War and leaned towards the left. These were "more ecumenically atheist. They do not even believe in science as an expression of religious yearning ... The fact remains that the parties do exist, and that Stephen Jay Gould and Richard Dawkins are not only their most visible proponents but also essential to defining them ... Both sides claim to be the true heirs of Darwin." (Andrew Brown[24])

Gould and Dawkins agree on one point: that they are both the best writers in the business, the best popularizers of science. Dawkins' evaluation of Gould is that he actually does not know what he says; that if he knows what he says he does not mean it; and that his theory of

---

24. Andrew Brown: op. cit., pp. 50 ff, passim.

punctuated equilibrium is in fact purely Darwinian, such as Dawkins himself has been explaining it all the time. "Gould seems to be saying things that are more radical than they really are," Dawkins alleges. "He pretends. He sets up windmills to tilt at which aren't serious targets at all."[25] Elsewhere he explains: "The extreme Gouldian view – certainly the view inspired by his rhetoric, though it is hard to tell from his own words whether he literally holds it himself – is radically different from and utterly incompatible with the standard neo-Darwinian model. It also has implications which, once they are spelled out, anybody can see as absurd."[26]

The history of science, like history in general, is replete with ardent polemics, in some of which even the greatest scientists were involved. The research of entire lives, the esteem of colleagues or the public, and many careers have depended on them. The quotes in the previous paragraph will give an idea of the bluntness of the controversy, and of Dawkins' "assurance in ridiculing [Gould's] ideas," as he states in *Unweaving the Rainbow*, which contains one of his main attacks on the American. The quotes are also chosen to show the level on which "the scientific battle for the soul of man" was and is being waged.

The sudden "macromutations" necessary for the macroevolution of the punctuated equilibrium theory remained unexplained. This was an occasion for another of Dawkins' attacks on Gould: the big jumps, required for a macromutation and the sudden appearance of new organs or whole species, were simply accumulations of the small steps required by Darwinism. Consequently punctuated equilibrium was nothing new and little more than a publicity stunt. "What needs to be said now, loud and clear," Dawkins wrote, "is the truth that the theory of punctuated equilibrium lies firmly within the neo-Darwinian synthesis. It always did."[27]

But knowing the ambiguity of Dawkins' reasonings, one will not be surprised to find him writing some years later that the Gouldian view was incompatible with the standard neo-Darwinian model. But then again he wrote: "A key feature of evolution is its gradualness ... There may be punctuations of rapid evolution, or even abrupt macromutations ... *Evolution is very possibly not, in actual fact, always gradual*. But it must

---

25. John Brockman: op. cit., p. 84.
26. Richard Dawkins: *Unweaving the Rainbow*, p. 202.
27. Richard Dawkins: *The Blind Watchmaker*, p. 311.

be gradual when it is being used to explain the coming into existence of complicated, apparently designed objects, like eyes."[28]

Behind the media hype about the genes, the genome and the wonders soon to be expected from the discoveries resulting from their study, many key problems in the science of biology remain unsolved. Most buzzwords, now as in the past, are only labels, pasted on empty jars. From the past we remember 'spontaneous generation', 'preformation', 'phlogiston'. A brief look will tell us whether 'gene', 'genome' or 'meme' belong in the same category, or not.

---

28. Richard Dawkins: *River out of Eden*, p. 97 (italics added).

11.

The Scientific Method

The single most characteristic feature of science is that its conclusions are always tentative, ready to be overthrown by new observational evidence and new theories that more compactly, more elegantly, and/or more completely explain the evidence.

JOHN L. CASTI

It is scientifically unsound to make assumptions of the way things ought to be.

MICHAEL BEHE

We have come a long way in our narrative of the various manners in which evolution was and is interpreted. A mass of facts which science has discovered since the time of the first geologists and paleontologists supports the view of a gradual development of the life forms on our planet. According to this view evolution is the only reasonably sustainable scheme and explanation of our biological past. What remains very much in question is the understanding of the way it all happened, closely dependent on the interpretation of the nature of matter, life and mind. Stepping aside for a short while from the information provided by our story of the evolutionary theories, a look at some essential points may clarify the total picture.

The Scientific Method

'The scientific method' is the much praised way of practicing science which is supposed to have changed a medieval world into a technological one, because it enabled humanity to unveil the secrets of nature and to use this newly acquired knowledge for mastering nature. The late Douglas Adams, author of *The Hitchhikers Guide to the Galaxy* and friend of Richard Dawkins, wrote: "The invention of the scientific method is, I'm sure we'll all agree, the most powerful intellectual idea, the most powerful intellectual framework for thinking and investigating and understanding and challenging the world around us that there is, and it rests on the premise that any idea is there to be attacked."[1]

The all-important first attack took place during the axis time of the Renaissance. The authoritarian dogmas and superstitions of the Catholic Church were put to the test of reason by intellectuals who were reading the rediscovered ancient Greek and Latin authors, and who, inspired by them, launched the movement they called *la nuova scienza*. A decisive confrontation in this general attack was the trial of Galileo Galilei by the Inquisition, lost by Galileo[2] but ultimately won by science.

As mentioned in one of the first chapters, Galileo's premises would become the foundations of the revolutionary scientific method. His first premise, and the most important one, was that only matter and material things should be the object of science. This created, from the start, a gap between science on the one side, and religion, occultism and everything else on the other. How sharply this separation was felt by both sides is illustrated precisely by the Galileo Affair, the core of which was the justification of different worldviews. Matter could be directly experienced by the senses; life and mind – and the non-material worlds of occultism, religion and fantasy or superstition conjured by them – could not. The fundamental materialism of science, and its offhand dumping of all else, have become so common that at present to academic science and its popularization the world is wholly and exclusively material.

Galileo's second premise was that science cannot handle wholes, it has to divide or reduce them into parts, consisting of smaller parts, consisting of still smaller parts: it is reductionist. This reductionism has been absolute in physical science since Galileo and Descartes, and was later on adopted by the biological sciences. A thing consists of parts; if we know all the parts and the parts of the parts, we know the whole. The 'mystic' view had always said the contrary, not that the parts explain the whole but that the whole explains the parts, and that the entire universe explains every single part of it, however large or little. "The "Universe is one. Its origin can only be the eternal unity. It is a vast organism in

1. Richard Dawkins: *A Devil's Chaplain*, p. 184.
2. Galileo Galilei is usually referred to as 'Galileo', his first name, which is quite exceptional. Nobody ever refers to Newton as 'Isaac' or to Einstein as 'Albert'.

which the natural things find their harmony and reciprocal sympathy." This view, called 'holism', is far from defunct and seems on the rise again, for a scientist of the stature of David Bohm has said: "It must always be remembered that, at a deeper level, attention must be given to the whole, which, in turn, acts to guide thought as it abstracts elements which do not in fact have a separate existence."[3] It is also this indelible intuition of the unity of all things that drives theoretical physics to continue its search for a Grand Unified Theory.

However, Richard Lewontin warns against "extreme holism" or "obscurantist holism." Even if it is true that "the whole is always prior to its parts" and everything is interconnected, "that should not be confused with the methodological claim that no success at all in understanding the world or manipulating it is possible if we cut it up in any way. Such a strong methodological claim we know to be wrong as a matter of historical experience. Whatever the faults of reductionism, we have accomplished a great deal by employing reductionism as a methodological strategy," wrote Lewontin recently.[4] It is, after all, reductionist science that has made our world. Yet the increasing awareness that "we need much more comprehensive [i.e. holistic] and much less reductionist understanding" may be a sign of "a new sort of science which is being forged at the moment." (Russell Stannard[5])

The problem of the whole and its parts is a problem of the mind, and hence cannot be solved by science itself as it recognizes only matter. Science is an exercise of the mind, a mental activity – a blatant truth which is often negated or overlooked. "Mind establishes this fiction of its ordinary commerce that [the given objects] are things with which it can deal separately and not merely as aspects of a whole," writes Sri Aurobindo. "For, even when it knows that they are not things in themselves, it is obliged to deal with them as if they were things in themselves, otherwise it could not subject them to its own characteristic activity."[6] This means that "a new sort of science" would have to be the child of a different kind of mind, one that can work with wholes and is not forced, because of its own constitution, to cut everything into parts.

Galileo's third premise is that all changes in matter are brought about by external forces. Matter is dead, it has no internal life or internal

---

3. David Bohm and David Peat: *Science, Order and Creativity*, p. 142.
4. Richard Lewontin: *The Triple Helix*, p. 110.
5. Russell Stannard: *Science and Wonders*, p. 172.
6. Sri Aurobindo: *The Life Divine*, p. 162.

dispositions to react. Moreover, internal actions and reactions, like those in animals and humans, cannot be determined quantitatively, they cannot be measured and represented by mathematical formulas. This tenet of the scientific method excludes from its field of examination an enormous part of phenomena essential to living organisms and to life itself. It leads unavoidably to the view that the whole of all living beings consists of material elements, which react to each other through external forces. The prime example of this view is the Cartesian metaphor of the machine.

Nothing in the universe exists by itself; everything is hierarchically interconnected with larger entities, to which it belongs, and with smaller entities, which are part of it. (Arthur Koestler gave the name "holon" to a thing in this multi-relationship.[7]) As the mind cannot grasp the totality even of the simplest thing in existence, it projects on it a simplification which makes the thing determinable and perhaps reconstructable. The metaphor of the machine is the consequence of the limited capacities of the mind.

A simple machine is evidently an artifact, but scientific materialism supposes that an ever increasing physical complexity will, at some undefined point, suddenly turn into a living organism. "The entire body of modern science rests on Descartes' metaphor of the world as a machine, which he introduced in Part V of the *Discourse on Method* as a way of understanding organisms, but then generalized as a way of thinking about the entire universe," writes Lewontin.[8] In Descartes' days automata that could perform amazing feats like gesturing, rolling their eyes or whistling, were all the rage. "Wandering through the Royal Gardens, Descartes was impressed by some water-driven robots and theorized that human and animal action was likewise a machine-like 'reflex'."[9]

"While we cannot dispense with metaphors in thinking about nature," continues Lewontin, "there is a great risk of confusing the metaphor with the thing of real interest. We cease to see the world *as if* it were *like* a machine and take it to *be* a machine. The result is that the properties we ascribe to our object of interest and the questions we ask about it reinforce the original metaphorical image, and we miss the

---

7. "The organism is not a mosaic aggregate of elementary physico-chemical processes, but a hierarchy in which each member, from the sub-cellar level upward, is a closely integrated structure." Arthur Koestler: *The Ghost in the Machine*, p. 64.
8. Richard Lewontin: op. cit., p. 3.
9. Larry Witham: *By Design*, p. 192.

aspects of the system that do not fit the metaphorical approximation."[10] This is how the cell came to be called a machine (by Monod) and that animals, including humans, are called robots (by Dawkins). There are futurists who expect man-made robots, within half a century or so, to be living organisms.

All this follows from the third premise that in science only external forces are legitimate, at first sight an innocent statement but deadening in its effects. "The problem for biology is that the model of physics, held up as the paradigm for science, is not applicable because the analogues of mass, velocity, and distance do not exist for organisms. ... Organisms move in a viscous medium; they suffer friction; they are too small and too distant from each other to interact gravitationally; their collisions are not elastic; their shapes, masses, and centers of gravity are changing; if they live in water they are buoyant; their paths are constantly being influenced by *external and internal* forces. The characteristic of a living object is that it *reacts* to external stimuli rather than being passively propelled by them," writes Lewontin, who is a geneticist at Harvard University.[11]

Science, as held by Galileo's fourth premise, can only work with the primary qualities of things: extension, motion, and mass. Secondary qualities like colour, scent and taste are conditioned by the primary qualities. Like all elaborate mental formations, the scientific method is an instrument, a set of mental formulations which fit more or less together as a whole, and which apply to a certain aspect of 'objective' reality. The primary qualities aforementioned are the ones the scientific method can handle. The secondary qualities, sometimes called 'qualia', escape its examining grip and are therefore considered of minor importance. Again, the division in primary and secondary qualities was necessary to enable any science to be done at all. But, again, this division has impoverished the world in which we live, reducing it as it were to black and white, this in total contradiction with our experience.

Galileo's fifth premise says that the language of science is mathematics, using the data of measurement. To quote his own words from *Il Saggiatore*: "Philosophy is written in that vast book which stands forever open before our eyes, I mean the universe; but it cannot be read until we have learnt the language and become familiar with the characters in which it is written. It is written in mathematical language, and the

---

10. Larry Witham: op. cit., p. 4 (italics in the text).
11. id., p. 93 (italics added).

letters are triangles, circles and other geometrical figures, without which means it is humanly impossible to comprehend a single word."[12] The importance of this step can only be realized when weighing it against the science of the medieval scholastics, which was mainly rhetorical verbosity and sophistry, a mental juggling game in the void with quotations from ancient authors and references as "proofs" from the Bible, the Church Fathers, and dozens of other mostly contradictory sources.

The last premise had perhaps the most direct influence on the founding of *la nuova scienza*: all guesses, theses, or theories have to be tested as to their truth and reality. It was the authentic, innate need of truth that, after centuries of theological, philosophical and pseudo-scientific fiction, led to the rule of the experiment as an absolute precondition for the acceptance of any idea, thesis or theory. "Correctness is more likely to be obtained by the experimental method than by any other process," writes Lewontin.[13] Richard Feynman put it as follows: "In general we look for a new law by the following process. First you guess. Don't laugh, this is the most important step. Then you compute the consequences. Compare the consequences to experience. If it disagrees with experience, the guess is wrong. In that simple statement is the key to science. It doesn't matter how beautiful your guess is or how smart you are or what your name is: if it disagrees with experience, it's wrong. That's all there is to it."[14]

Experimentation became the norm during the birth period of the modern sciences. William Gilbert (1544-1613), in his great book on magnetism *De Magnete*, was one of the first "to set out clearly in print the essence of the scientific method: the testing of hypotheses by rigorous experiments." Realdus Columbus, Vesalius' successor as professor of anatomy in Padua, told his students: "Try the experiment and find out whether what I have said agrees with the thing itself." (Formerly, anatomy lessons consisted mainly in reading passages from Galen as comments by dissections.) William Harvey (1578-1657), who discovered the circulation of the blood, wrote: "I do not profess to learn and teach anatomy from the axioms of the philosophers, but [directly] from dissections and from the fabrick of nature." This experimental approach was enthusiastically espoused by the Royal Society, and

---

12. P.H. Crombie: *Medieval and Early Modern Science*, vol. II, p. 142.
13. Ullica Segerstråle: *Defenders of the Truth*, p. 105.
14. Simon Singh: *Big Bang*, p. 357.

firmly established, well before the end of the seventeenth century, as *the* scientific method."[15]

Is There a Scientific Method?

To ask this question after the matter discussed in the previous section may seem nonsensical. Yet, Isabelle Stengers sounds a first warning: "Every science has its own methods which cannot be applied without precautions to other sciences. Moreover, the methods evolve within the same science. To speak of *the* experimental method of physics or biology is omitting to take into account the evolution of the practices and the kinds of argumentation proper to each particular period, country, or even [scientific] institution."[16] Henry Bauer, in his *Scientific Literacy and the Myth of the Scientific Method*, goes at it more directly: "The scientific method is a myth, it does not explain the success of science, and scientists in practice do not follow the method."[17]

Science, writes Bauer, "begins by chance and caprice, at the frontier, with hardly a shadow of the scientific method in evidence," after which it is "sieved, tested and modified until it appears in the textbooks." This sieving and testing happens in the course of the procedure of trying to have a new idea or discovery accepted by the community of scientists. First there is the writing of a paper expounding the idea; then the paper must be submitted to one of the numerous publications in the discipline, where it is reviewed by a jury of senior scientists called 'peers'; if the peers find the idea interesting and the authors of the paper are lucky, it will be published. More luck may bring the paper – one in a stream of hundreds which are continually published – to the attention of the scientific community. If the new idea is accepted, it may be integrated into the discipline's reigning paradigm and perhaps in the general scientific paradigm. After having negotiated all these hurdles, the new idea may be included in the text books, the ultimate consecration, and spread through academia.

"The ubiquity of intuition in science can scarcely be overlooked," writes John Ziman, confirming that "science begins by chance and

---

15. John Gribbin: *The Fellowship*, pp. 3, 104. 119, 50.
16. Isabelle Stengers and B. Bensaude-Vincent: *100 mots pour penser la science*, p. 244 (italics in the text).
17. Henry Bauer: *Scientific Literacy and the Myth of the Scientific Method*, p. 39.

caprice."[18] It is true that in most cases the intuition is the spark made possible by the tension of much reflection or experimenting. Newton's apple is a well-known example; whether a real event or legend, it certainly tells of an instant illumination. We have seen how Wallace's theory of evolution came to him "in a sudden flash of insight" during a bout of fever on a tropical island. August Kekulé 'saw' the structure of the benzene ring, the key to organic chemistry, while dozing. Descartes had his three dreams in a similar condition; they were the call to his vocation and the occult instigation of his philosophy of rationalism. Mendeleyev's fundamental insight of the periodic table of elements came also to him in a dream.

Albert Einstein narrated: "The breakthrough came suddenly one day. I was sitting in a chair in my patent office in Bern. Suddenly the thought struck me: If a man falls freely, he does not feel his own weight. I was taken aback. This simple thought experiment made a deep impression on me. This led me to the theory of gravity."[19] For Watson and Crick, co-discoverers of the double helix structure of the DNA molecule, "the penny dropped, a moment of great insight." "At three o'clock one morning, lying sleeplessly on his bed in a small hostel, Werner Heisenberg knew that he had the tool enabling him to perform calculations in his new [quantum] mechanics. So he rose from his bed and started figuring. In his feverish state he made endless slips and errors and had to start over again and again. But finally he got an answer, and it was more than he could have dreamed for. What he had found was a gift from above, he thought, a discovery of unwarranted and unexpected proportions."[20]

These examples must suffice, but it seems that hardly any important theoretical discovery is the end product of an effort of logical reasoning: most discoveries, and certainly the important ones, are the consequence of a sudden illumination. The same conclusion could be drawn from the history of biological and technological research. The way Alexander Fleming discovered penicillin will readily come to mind, but Royston Roberts has filled a volume with "accidental discoveries in science" and called it *Serendipity*. Scientism attacks with disdain all forms of irrationality on every possible occasion. Yet, strange to say, it has no explanation for the phenomenon of rational consciousness, which it supposes to be a material "activity of neurons in the central

18. John Ziman: *Reliable Knowledge*, p. 101.
19. Marcus Chown: *Quantum Theory Cannot Hurt You*, p. 117.
20. David Lindley: *Uncertainty*, pp. 113-14.

nervous system;" and the history of science is a succession of irrational illuminations without which neither science nor scientism would exist.

"What has been presented as the scientific method, at any given time, has been a simplified snapshot of an intrinsically much more opportunistic enterprise," writes Piet Hut, the Dutch astronomer. "The strength of science is not at all in its currently accepted method. The strength is the fact that scientists allow the method to change."[21] And Lewis Wolpert concludes his enquiry among fellow scientists thus: "Many famous scientists have given advice [to their younger colleagues]: try many things; do what makes your heart leap; think big; dare to explore where there is no light; challenge expectation; *cherchez le paradoxe*; be sloppy so that something unexpected happens, but not so sloppy that you can't tell what happened; turn it on its head; never try to solve a problem until you can guess the answer; precision encourages the imagination; seek simplicity; seek beauty … One could do no better than to try them all. No one method, no paradigm, will capture the process of science. There is no such thing as *the* scientific method."[22]

The problem of the scientific method, especially in biology, becomes clearer if one realizes that the Galilean premises are relevant to *material* objects. Material objects have no internal reactions and are moved by impacts of external forces. If the external circumstances remain the same, a material objects will not change; therefore it can be counted, measured and weighed. But, as Lewontin remarks, first, in nature external circumstances never remain the same, and, second, "the characteristic of a living object is that it *reacts* to external stimuli rather than being passively propelled by them."[23] "Scientific laws only give a schematic account of material process of Nature – as a valid scheme they can be used for reproducing or extending at will a material process, but obviously they cannot give an account of the thing itself," writes Sri Aurobindo,[24] and certainly not of a living organism.

"… We live in the surface mind of ignorance, do not know what is going on behind and see only the phenomenal process of Nature. There the apparent fact is an overwhelming determinism of Nature and as our surface consciousness is part of that process, we are unable to see the other term of the biune reality. For practical purposes, on the surface there is an entire determinism in Matter – though this is now disputed

21. John Brockman (ed.): *What Are You Optimistic About?* p. 351.
22. Lewis Wolpert: *The Unnatural Nature of Science*, p. 108 (italics in the text).
23. Richard Lewontin: op. cit., p. 93 (italics added).
24. Sri Aurobindo: *Letters on Yoga*, p. 214.

by the latest school of Science [at the time of writing quantum mechanics]. As Life emerges a certain plasticity sets in, so that it is difficult to predict anything as exactly as one predicts material things that obey a rigid law."[25] And the plasticity increases with the growth of Mind.

Parading the Paradigm

While reading Aristotle's *Physics* Thomas Kuhn, a philosopher of science, also had his illumination. "During that moment Kuhn saw – he knew! – that reality is ultimately unknowable; any attempt to describe it obscures as much as it illuminates."[26] The result of this 'epiphany' was *The Structure of Scientific Revolutions*, first published in 1962. According to Steve Fuller it is "the most influential book on the nature of science in the second half of the 20th century – and arguably, the entire 20th century."[27] John Horgan expresses the opinion that it "may be the most influential treatise ever written on how science does (or does not) proceed."[28]

It was this revolutionary book which caused the indiscriminate use of the buzzword 'paradigm', applied as well to the political strategy of a presidential candidate as to a new formation of a football team. "A paradigm is an accepted model or pattern," writes Kuhn, making it at first look quite simple and innocent. As he further defines the term, paradigms are "universally recognized scientific achievements that for a time provide model problems and solutions to a community of practitioners ... By choosing this term, I mean to suggest that some accepted examples of actual scientific practice provide models from which spring particular coherent traditions of scientific research. ... A paradigm is what the members of a scientific community share, and, conversely, a scientific community consists of men who share a paradigm."[29]

In science, a paradigm is the commonly accepted interpretation of the universe, nature or reality by the practitioners of a scientific discipline. For instance, there is the Aristotelian view of the universe and there is the Newtonian view – both very different from the view of the ancient Chinese or the American Indians, who had their own

---

25. id., p. 474.
26. John Horgan: *The End of Science*, p. 47.
27. Steve Fuller: *Kuhn vs Popper*, p. 18.
28. John Horgan: op. cit., p. 41.
29. Thomas Kuhn: *The Structure of Scientific Revolutions*, pp. 23, x, 10, 176.

kind of science. To Aristotle the Earth was the centre of the universe and all heavenly bodies turned around it. The Earth was stable but corruptible in its elements, the heavenly bodies were neither. Aristotle had an explanation for all phenomena perceived in that universe (the planets moved on crystalline spheres pushed forward by angels), and his model was the standard and even obligatory one in the Western world till Copernicus taught differently. To Isaac Newton the Earth was no longer the centre of the universe. His calculations proved that the solar system functioned like a clockwork mechanism, and that it was gravity which held the system together and moved the planets. What gravity was remained a mystery, but the calculations fitted the perceived movements of the bodies. Newtonian physics were universally recognized till replaced by the paradigm of Einsteinian relativity.

The substitution of the Aristotelian worldview by the Copernican was the background of the controversy between the Catholic Church and Galileo. Newton's clockwork universe was violently opposed, one of the reasons being the force of gravity which, like magic, worked at a distance. All new paradigms (including Kuhn's) have been systematically opposed by the proponents of the established ones. Why? Because a new paradigm has to conquer the minds and hearts of the scientific community and its acceptance is as traumatic as a religious or political conversion. "The transfer of allegiance from paradigm to paradigm is a conversion experience."[30] The reigning paradigm, during the time of its validity, is *the truth*; a different idea puts this truth into question and is therefore considered rebellious and unorthodox, or absurd, or plain stupid. In these matters science at times does not differ from religion, but is as dogmatic and aggressive – and so are the fundamental political systems, a fact abundantly illustrated by the history of communism. To accept a new paradigm "the scientist's perception of his environment must be re-educated,"[31] he must learn to see the same facts differently, and to perceive new facts which he did not perceive before although they were there.

Essential to the understanding of these attitudes is the working of the human mind and its activities, be they scientific, religious, or political. The function of the mind "is to cut out something vaguely from the unknown thing in itself and call this measurement or delimitation of it the whole, and again to analyze the whole into its parts which it

---

30. id., p. 10.
31. Thomas Kuhn: op. cit., p. 112.

regards as separate mental objects. It is only the parts and accidents that the Mind can see definitely and, after its own fashion, know."[32] The construction of a new paradigm demands a huge, life-changing mental effort, putting together a considerable number of newly interpreted facts discovered by reflection, calculation and research. To arrive at accepting such a new mental formation – the process of conversion – takes time and demands an effort which cannot be repeated in one human life. For scientists, this effort means also the work of a lifetime, abandoning which would mean erasing years of work, prestige and self-esteem. This is one of the reasons why a given paradigm is always defended with all possible means and even against reasonable arguments.

It is also the explanation of the fact that the practitioners of a scientific specialty must be "professionally initiated." "Professionalization leads, on the one hand, to an immense restriction of the scientist's vision and to a considerable resistance to paradigm change. The science has become increasingly rigid."[33] Science becomes as dogmatic as a religious church or a totalitarian political system. It becomes irrational. "By losing its hypothetical character to become an unshakable dogma, the dominant theory [i.e. the paradigm] takes on the aspect of a myth."[34] "Is there any defence against the charge that the whole scientific paradigm is a self-sustained delusion?" asks John Ziman. "The scientists in our model are almost always deliberately trained to a particular attitude to natural phenomena. How are their intellectual constructs to be distinguished from those of any other self-accrediting social group, such as a religious sect? What reason have we for preferring the scientific paradigm as the ideal, unique world picture?" The institutions where scientists are educated and trained provide "the brainwashing implicit in the long process of becoming technically expert."[35] They are no longer temples of science, but of scientism.

All this may explain why Thomas Kuhn's essay on the formation of scientific paradigms and their eventual replacement was so vigorously opposed. Science, in his view, is not constant progress towards the discovery of the ultimate truth; it consists of islands of partial truth, held to be absolute for a limited period, after which, in a kind of mental volcanic eruptions, they will disappear and new islands will be formed, perhaps to join a small archipelago of still surviving ones. "Theory

---

32. Sri Aurobindo: *The Life Divine*, p. 128.
33. Thomas Kuhn: op. cit., p. 64.
34. Gerard Amzallag: *La raison malmenée*, p. 132.
35. John Ziman: op. cit., p. 8

is preconceived belief," writes Henry Bauer.[36] He could have written "hardened belief."

A fierce controversy developed between Kuhn, the other prominent philosopher of science Karl Popper, and their adherents. For Popper there *was* an absolute Truth or Reality which science step by step discovered, although the correctness of scientific formulas could never be proved but only disproved or "falsified." For Kuhn, even if there is an absolute Reality, it can never be known by science; only mental conceptions, artificially composed images of it, are what science is able to accomplish. The root of the problem is, again, the human mind. "We regard thought as a thing separate from existence, abstract, unsubstantial, different from reality, something which appears one knows not whence and detaches itself from objective reality in order to observe, understand and judge it; for so it seems and therefore is to our all-dividing, all-analyzing mentality. The first business of mind is to render 'discrete', to make fissures much more than to discern, and so it has made this paralyzing fissure between thought and reality." (Sri Aurobindo[37])

The impact of Kuhn's insight on the understanding of the role of science has been enormous. It is the cause why at present one reads statements like the following by Owen Gingerich, professor of astronomy and of the history of science: "Today physics marches on not so much via proofs as through the persuasive coherency of the picture it presents. What passes for truth in science is a comprehensive pattern of interconnected answers to questions posed to nature – explanations of *how* things work (efficient causes), though not necessarily *why* they work (final causes)."[38] The certitudes of scientism have become relative.

Biology and Physics

If there is a date which may be pinpointed as the start of the modern scientific revolution, it is 1543. This date will be readily recognized as the year of the publication of Nicolaus Copernicus' *De Revolutionibus Orbium Celestium*. It is less commonly known that, by a wonderful coincidence, it is also the publication date of the magnificently illustrated

36. Henry Bauer: op. cit., p. 23.
37. Sri Aurobindo: op. cit., p. 130.
38. Owen Gingerich: *God's Universe*, p. 95.

book *De Humani Corporis Fabrica* (about how the human body is made) by the Flemish anatomist Andreas Vesalius.[39] Thus the gates of two parallel terrains of scientific exploration opened at the same time, but physics, supported by mathematics, would advance like a speedy hare while the biological sciences, in comparison, would be the proverbial tortoise.

As just narrated, physics had invented the trick to reduce all things under its consideration to matter, to dead objects. Biology found itself unable to exert a similar strong and clearly definable grasp on the subjects under its consideration, which were living organisms. A dead bird does not move unless moved, a living bird flies away of itself. In the 1660s, when the main works of Galileo, Kepler and Descartes had been published and Newton was already working on his grand synthesis, "European medical knowledge and teaching were in a state of flux. New anatomical and physiological discoveries such as the circulation of the blood had undermined confidence in the tradition of Aristotle and Galen without necessarily replacing it with anything more coherent or effective. ... Medicine remained in a pre-scientific state ..."[40]

Unable to invent a method to study living organism scientifically, biologists of all disciplines turned toward the successful sister-science: physics. "Emerging more than a century after modern physics, biology saw itself challenged by the scientific criteria of its elder sister. On the one hand the biologists could profit from the experience of their colleagues in physics; on the other the presence of a science in full expansion invited the biologists to imitate its methods, supposing that they would warrant the success of their own enterprise."[41]

"A key ambition of the scientific revolution was to provide numerical, objective descriptions of all aspects of the universe, including living and anatomical phenomena," writes Matthew Cobb. He quotes the Italian anatomist Francesco Redi: "I wished to demonstrate in these dissertations that unless myology [the study of the muscles] becomes part of mathematics, the parts of the muscles cannot be distinctly designated nor can their movement be successfully studied."[42]

The efforts by the biologists to go beyond Aristotle and Galen, the two medieval authorities on all things biological, were admirable. Most

---

39. Much of the seventeenth century scientific revolution was prefigured in the notebooks of Leonardo da Vinci, but they remained for the most part unknown.
40. Matthew Cobb: *The Egg & Sperm Race*, pp. 42-3.
41. Gerard Amzallag: op. cit., p. 51.
42. Matthew Cobb: op. cit., p. 95.

of their names have been forgotten, although they deserve better. Niels Steno, Reinier de Graaf, Jan Swammerdam, Johannes van Horne, and many others, were masters who travelled through Europe, and studied and taught in all high places of medicine: Leiden, Padua, Montpellier, Paris. To them we own much of the knowledge of our own bodies and those of our evolutionary relatives. Antonie van Leeuwenhoek, in continuous correspondence with the newly founded Royal Society, demonstrated the wonders of the microscope. The process of animal procreation was discovered. Life and its miracles seemed within the reach of understanding and imitation.

Common to all of them was the new scientific method of the experiment and a real enthusiasm for it. The new method was based on rationalism, logic, and scepticism towards all traditional theories, in most cases not better than invention and fancy. From that time onwards biology suffered from what has been called 'physics envy', caused by the fact that physics progressed on the basis of rational proof and the logic of mathematics, while biology remained confronted with the complexity of the organism, its characteristics of reproduction and alimentation, and behind it all the mysterious, scientifically untreatable force of life. Vitalism, the acceptance of the reality of such a life force, kept and keeps raising its head. Around 1900 it became, with figures like Henri Bergson and Hans Driesch, a widespread and outspoken movement. It has systematically been condemned by academic biologists as mysticism, animism or occultism.

Richard Dawkins proclaims: "Everything ultimately obeys the laws of physics."[43] We know of the ambition of his fellow sociobiologist Edward Wilson to make biology into the one foundation and explanation of the universe, his kind of Grand Unified Theory. "Wilson's sociobiology is an attempt to make evolutionary biology a total quantitative and predictive science." (Ullica Segerstråle[44]) "Biology is nothing but chemistry, and chemistry is nothing but physics," states Robin Dunbar.[45] Francis Crick agrees: "It is indeed the goal of modern biological research to make the whole of biology understandable in the way physics and chemistry are formulated."[46] This results in the following sentences from a recent book to teach biology: "The living cells with which biology is concerned can be reduced to the chemical compounds of which

43. Richard Dawkins: *The God Delusion*, p. 181.
44. Alan Grafen and Mark Ridley (ed.): *Richard Dawkins*, p. 82.
45. Robin Dunbar: *The Trouble with Science*, p. 88.
46. John Lennox: *Hat die Wissenschaft Gott begraben?* p. 34.

they are made and then further reduced to the constituent atoms, which follow the laws of physics. Thus, in this approach, although the human body is complex, its operations can ultimately be analyzed in terms of the universal laws of physics."[47]

What seems to have escaped 'orthodox' present-day biology, however, is that the physics it envies is that of the nineteenth century, and that since Einstein and especially since the birth of quantum mechanics physics has gone off in very different directions. "So we are now in the very strange position that whereas physicists are implying that, fundamentally and in totality, inanimate matter is not mechanical, molecular biologists are saying that whenever matter is organized so as to be alive, it is completely mechanical," reflects Mary Midgley, a philosopher who made her voice heard in the sociobiological debate.[48] "It is one of the greatest ironies of our century," wrote Michael Talbot in the 1990s, "that as reductionist biologists were slowly trying to purge all mention of consciousness from their understanding of neurophysiological processes, physicians were at the same time uncovering compelling evidence that the mind is not only necessary, but may be integral to our understanding of the physical universe."[49]

"The problem for biology is that the model of physics, held up as the paradigm for science, is not applicable because the analogies of mass, velocity and distance do not exist for organisms." (Lewontin[50]) The classical machine model fails in biology. A material molecule cannot become alive or conscious; material parts joined together in whatever way do not become a living organism. In physics causal claims are generally based on the precondition that all elements of the experimental system remain equal, "but in biology all other things are almost never equal." (Lewontin) Each snowflake has a configuration that depends on its distinctive history. And we recall Lewontin's words, that "the characteristic of a living being is that it reacts to external stimuli rather than being passively propelled by them."

In this context the terse statement of Ernst Mayr, a founder of neo-Darwinism, deserves to be mentioned: "All so-called evolutionary laws are contingent [i.e. accidental, random] generalizations ... with numerous exceptions, and are quite different from the universal laws of physics." And he quotes Bernhard Rensch, another architect

---

47. *Teach Yourself Biology*, p. 146.
48. Mary Midgley: *Evolution as Religion*, p. 115.
49. Michael Talbot: *Beyond the Quantum*, p. 149.
50. Richard Lewontin: op. cit., p. 93.

of neo-Darwinism: "Evolutionary 'laws' are greatly restricted in time and place and therefore do not satisfy the traditional definitions of scientific laws."[51] So we find, firstly, that physics and biology are quite different fields of scientific study and experimentation which call for their own kind of laws; secondly, that evolution, the general object of our exploration, is *historical*, a sequence of one-time events, each of them particular and never exactly repeated. A law cannot be the general abstract description of single, one-time events – which is the error at the basis of Lyell's and Darwin's gradualism, stubbornly defended by Dawkins.

"There is no reason to think that the laws of physics are violated in living matter. There is nothing supernatural, no 'life force' to rival the fundamental forces of physics. ... The body is a complex thing with many constituent parts, and to understand its behaviour you must apply the laws of physics to its parts, not to the whole. The behaviour of the body as a whole will then emerge [?] as a consequence of interaction of the parts. ... It is only when we remember that it has many parts, all obeying laws of physics at their own level, that we understand the behaviour of the whole body. This is not, of course, a peculiarity of living things. It applies to all man-made machines, and potentially applies to any complex, many-parted object." Thus goes the gospel of arch-reductionist Richard Dawkins in his book *The Blind Watchmaker*.

Then, a few pages further on, we read: "I am a biologist. I take the facts of physics, the facts of the world of simplicity [?], for granted. If physicists still don't agree over whether those simple facts are yet understood, that is not my problem. My task is to explain elephants, and the world of complex things, *in terms that physicists either understand, or are working on*. The physicist's problem is the problem of ultimate origins and ultimate natural laws. The biologist's problem is the problem of complexity. The biologist tries to explain the working, and the coming into existence, of complex things, in terms of simpler things. He can regard his task as done when he has arrived at entities so simple that they can safely be handed over to physicists. ... The kind of explanation we come up with must not contradict the laws of physics. Indeed it will make use of the laws of physics, and nothing more than the law of physics."[52]

In these passages physics is accepted as the ultimate foundation of

---

51. Ernst Mayr: *What Evolution is*, p. 251.
52. Richard Dawkins: *The Blind Watchmaker*, pp. 13-4, 19, 20 (italics added).

biology. Yet, the "simple facts" of physics, which Dawkins takes for granted, seem not yet to be understood. Then how can he "safely hand over to the physicists" the fruits of his reductions, and on what basis are his reductions made? – Are the foundations of physics still in doubt? The answer seems to be: now more than ever, as discussed in books like John Horgan's *The End of Science,* Lee Smolin's *The Trouble with Phycics,* or Peter Woit's *Not Even Wrong.* There is indeed no reason to think "that the laws of physics are violated in living matter," as the research beyond the circle of Dawkins' thought tells us that they are *not applicable* to living matter in the way they are to dead matter.[53]

"It is not new principles that we need," writes Lewontin, "but a willingness to accept the consequences of the fact that biological systems occupy a different region of the space of physical relations than do simpler physico-chemical systems, a region in which the objects are characterized, first, by a very great internal physical and chemical heterogeneity and, second, by a dynamic exchange between processes internal to the objects and the world outside them. That is, organisms are internally heterogeneous open systems."[54] It may be remembered that the author of these words is a geneticist and professor of biology at Harvard. Steven Rose, another professor of biology, at the University of London, said in an interview: "There is a strand within physics which says that physics is the ultimate science, and that everything else has to reduce to it. One of the things I am concerned with as a biologist is to say that physics is a model for doing *physics*; it is not a model for understanding the biological world. We [biologists] deal with much more complicated phenomena than physics. The brain is, I suspect, the most complex organization of matter in the universe. That degree of complexity doesn't reduce to very simplistic models."[55]

"Systems biology is where we are moving to," writes Denis Noble, formerly professor at Oxford. "Only, it requires a quite different mind-set. It is about putting together rather than taking apart, integration rather than reduction. It starts with what we have learned from the reductionist approach; and then it goes further. It requires that we develop ways

---

53. "Many scientific theories refer to unverifiable and unobservable entities such as forces, fields, molecules, quarks, and universal laws. ... Forces, fields atoms, quarks, past events, mental states, subsurface geological features, molecular biological structures – all are unobservables inferred from observable phenomena. Nevertheless, most are unambiguously the result of scientific inquiry." (Behe, Dembski and Meyer: *Science and Evidence for Design in the Universe,* pp. 158, 169.)
54. Richard Lewontin: *The Triple Helix,* p. 114.
55. Russell Stannard: *Science and Wonders,* p. 171 (italics in the text).

of thinking about integration that are as rigorous as our reductionist procedures, but different. This is a major change. It has implications beyond the purely scientific. It means changing our philosophy, in the full sense of the term."[56]

---

56. Denis Noble: *The Music of Life*, p. xi.

12.

Of Genes, Genetics and Genomes

There's always a danger that people think that because you have a
Nobel Prize in something, you know something about other things.

WILLIAM PHILLIPS (Nobel Prize in Physics 1997)

From Complexity to Perplexity

Physically speaking, genes are sequences of the DNA molecules which constitute the chromosomes in a cell's nucleus. Functionally speaking, genes are the units that cause the existence and determine the formation of all living organisms. Anyway, this is what the textbooks presently in use are teaching. We are familiar with the history of the hereditary factors which has led to the present concept of the gene. We have met with Darwin's gemmules, Weismann's germ plasm and biophores, de Vries' pangenes and their mutations, the supposed locations (*loci*) of certain genes on the chromosomes, and the discovery of the structure of DNA molecules, the physical elements of the genes.

We also know of "the Weismann barrier," the thesis which said that evolution is a process solely going from the genes and the genotype to the materialized organism, the phenotype, never the other way round. This principle stresses the primary importance of the genes, and is closely related to Francis Crick's "central dogma", namely that "DNA makes RNA makes protein," never inversely. "Genetic information flows in only one direction: from DNA outwards. The statement is called 'the central dogma' of molecular genetics. It has been elaborated from a vast array of experimental data and seems unlikely ever to be seriously challenged."[1] Crick, in the opinion of John Horgan one of

---

1. Robert Shapiro: *Origins*, p. 290.

the most ruthless reductionists in the history of science, "desired to show that life was indeed mechanistic."[2]

Such was the way from Darwin's ignorance of the hereditary process to the triumph of the genes, called "DNA mysticism" and "DNA mania" by André Pichot. The triumphalism of the genes reached its zenith with Richard Dawkins in his book *The Selfish Gene*: "The argument of this book is that we, and all other animals, are machines created by our genes." All life on Earth and in the universe, Dawkins proclaims, is the work of the genes, which for tactical reasons he started calling "replicators." The origin, the *fons et origo* of life was not longer to be sought in God, a superior Intelligent Power, or in whatever. It was now located in the genes and their concerted action in the gene pools, for reasons never explained.

Around 1950 biochemical research in nuclear acids was still looked down upon as of little importance. This made Watson and Crick's discovery of the double helix all the more sensational. "For those not studying biology at the time in the early 1950s," said Edward Wilson, "it is hard to imagine the impact the discovery of the structure of DNA had on our perception of how the world works."[3] Heredity, and consequently evolution, were no longer a matter of fictitious biophores or the calculation of probabilities, but a concrete material structure and mechanism. "Watson and Crick's achievement stands unrivalled in the annals of twentieth-century biology," asserts Evelyn Fox Keller.[4] As a result, biochemistry took on a sudden surge. Between 1953 and 1963 were established: the mechanism of DNA duplication, the existence and the role of messenger RNA and transfer RNA, the genetic code, the mechanism of protein synthesis, and the general principles of the regulation of this synthesis. From this followed practically everything biological which is the standard fare in the media nowadays, such as genomes, clones, microbiology, nanotechnology, and the prediction that in 2050 designer babies will be among us.

---

2. Larry Witham: *The Measure of God*, p. 232. "The dogma itself was first pronounced by Francis Crick in 1958. In its exact words, it stated: 'The transfer of information from nucleic acid to nucleic acid, or from nucleic acid to protein may be possible, but transfer from protein to protein, or from protein to nucleic acid is impossible.' ... Crick recalled that at one point after he had put forth his theory, a friend told him that a dogma is something that cannot possibly be doubted. 'I didn't know it meant that,' said Crick. 'I thought it meant a hypothesis, some arbitrary thing which was laid down for no particularly good reason. Otherwise it would have been called the Central Hypothesis ...'" (Robert Shapiro: *Origins*, p. 291)
3. Edward Larson: *Evolution*, p. 268.
4. Evelyn Fox Keller: *The Century of the Gene*, p. 13.

Mendel's statistics, de Vries' mutations and the mathematical theories applied to genetics may have been for some time relatively simple; defining in practice the exact functions of the huge strings of genes proved exceedingly complex. This was one of the reasons why it was accepted in principle that each hereditary character was brought about by *one* gene, sometimes called the "one-to-one" theory. Ernst Mayr writes: "For the sake of simplicity, it was traditionally assumed that each gene acted independently of all others. ... Scientists assumed that the same gene, no matter where found, always had the same phenotypic effect."[5] Another reason was that the one-on-one thesis suited the machine model in biology nicely. "The living being was thus cut up in hereditary characteristics of which each one was associated to a gene, in other words a particle of heredity, whose existence was postulated, but of which one knew neither the nature nor the effect."[6]

This is the way genetics is still commonly understood. Day after day the discovery of a new correlation between a certain gene and a phenotypic characteristic is announced in the media. "Never in the history of the gene has the term had as much force in the popular imagination as in recent years," writes Fox Keller, "and, accordingly, never has gene talk had more persuasive – that is, rhetorical – power ... The image of genes as clear and distinct causal agents, constituting the basis of all aspects of organismic life, has become so deeply embedded in both popular and scientific thought that it will take far more than good intentions, diligence, or conceptual critique to dislodge it. So, too, the image of a genetic program – although of more recent vintage – has by now become equally embedded in our ways of thinking."[7] In simpler words: certain biological myths about genes and genomes are being enforced on the public mind by scientifically unfounded repetition in the media.

A few examples of the one-to-one theory in widely read science magazines must suffice. "Divorce gene blamed for strained marital relation ... mapping the cancer genome ... sensitivity to bad odours due to a gene ...unlocking the secrets of the longevity genes ... alcoholism and our genes ... the gene which makes meat tender has been identified ...which genes cause heart disease ... finding mental illness genes ..." "Over the past decade," writes John Horgan, "scientists have linked specific genes to manic depression, schizophrenia, autism, alcoholism, heroin addiction, high IQ, male homosexuality, sadness, extroversion,

5. Ernst Mayr: *What Evolution is*, pp. 118, 125.
6. André Pichot: *Histoire de la notion de gène*, p. 151.
7. Evelyn Fox Keller: op. cit., pp. 143, 136.

introversion, social skills, novelty seeking, impulsivity, attention-deficit disorder, obsessive-compulsive disorder, violent aggression, anxiety, seasonal effective disorder, etc. ... So far none of the claims linking specific genes to specific, complex behavioural traits and disorders – *not one* – has been unambiguously confirmed."[8]

Should the image of "genes as clear and distinct causal agents" then have to be dislodged from popular and scientific thought? Ernst Mayr – here quoted for his authoritative status in evolutionary biology – gives a warning signal. "The effects of genes on development are often surprisingly diverse. ... Almost every gene that has been studied in higher organisms has been found to effect more than one organ system, a multiple effect which is known as pleiotropy. ... It is doubtful whether any genes that are not pleiotropic exist in higher organisms."[9] Denis Noble's statement on this issue is radical: "There is no one-to-one correspondence between genes and biological functions. ... Many gene products, the proteins, must act together to generate biological functions at a higher level. ... Each gene may also play a role in many different functions ... To form a high level physical function, large numbers of genes are expressed simultaneously. Very probably as much as a third of the genome, 10 000 genes, may be expressed in an organ like the brain."[10]

(It maybe recalled, for the sake of clarity, that in the cell the main elements are the nucleus with its strings of genes, called chromosomes, and the different kinds of proteins. Proteins (RNA, ribonucleic acid) read the information stored in the genes and use it to build other proteins, which in their turn build the cells and organs that form the organism, and keep it alive. From 'lower' to 'higher' means here the hierarchic order of increasing complexity of the genes, proteins, cells, organs and the organism as a whole. The genome of an organism consists of all the genes contained in a single set of its chromosomes.)

The further the study of the genes progressed in the last decades, the more the complexity of the genes increased. Curious "jumping genes" had already been spotted by Barbara McClintock in the 1940s. A jumping gene, or transposon, is "a mobile genetic element that can become integrated at many different sites in the genome, either by moving from

---

8. John Horgan: *The Undiscovered Mind*, pp. 140-41 (italics in the text).
9. Michael Denton: *Evolution – A Theory in Crisis*, p. 149.
10. Denis Noble: *The Music of Life – Biology beyond the Genome*, pp. 9, 42.

place to place or by producing copies of itself that insert elsewhere in the genome."[11]

Then followed the discovery of "junk DNA", thus called because at first it seemed completely superfluous as it did not code for proteins. Junk DNA consisted of disabled genes or bits of genes scattered across the genomic landscape, and which had been there for millions of years. Astoundingly, more then 98 percent of the human genome seemed be such genetic trash, "stuff that does nothing." Dawkins soon had an explanation: "The simplest way to explain the surplus DNA is to suppose that it is a parasite, or at best a harmless but useless passenger [98 percent!], hitching a ride in the survival machines created by the other DNA."[12] Other conclusions were readily drawn, like: junk DNA showed that, if there was a Designer, "he had made serious errors, wasting millions of DNA molecules on a blueprint [of his designs or creatures] full of junk and scribbles."[13]

Yet, as Michael Behe noted, "it is scientifically unsound to make any assumption of the way things *ought* to be," things in this case being the genes and the genome. For it was gradually discovered that so-called junk DNA "contains regulatory sequences that tell other genes when to turn on and off and genes encoding RNA that does not get translated into a protein, as well as a lot of DNA having purposes scientists are only beginning to understand."[14]

(This is one of the examples of what might be called 'the reflex of simplification' humans show in the presence of all complex matters, also scientific ones. Drawings of the atom as a miniature solar system are still common – and always of the simplest one, the hydrogen atom – although the picture of the atom has drastically changed decades ago. DNA molecules are represented as minute sections of a double helix, while there number is astronomical and our bodies comprise trillions of cells, all alive and active. Molecules are represented as constructions of tiny spheres and half-spheres connected with tiny pipes, miniatures of the Brussels Atomium. For sure, making complete figures of atomic (and cosmic) realities is impossible, and representing them in simplified drawings is a necessity; but it should be kept in mind that this is a sign of our incapacity to 'imagine' the complexity of nature, which is a serious impediment to our understanding of it.)

---

11. *Oxford Dictionary of Biology*, p. 656.
12. Richard Dawkins: *The Selfish Gene*, p. 45.
13. Michael Behe: *Darwin's Black Box*, p. 226.
14. *Scientific American India*, May 2009, p. 36.

Another instance of the ever greater complexity molecular biologists are confronted with, is the folding of the proteins. Most proteins consist of some several thousand atoms folded into "an immensely complex spatial arrangement." "In ways we do not yet understand, the DNA threads are folded into a three-dimensional form in the nucleus of each cell. ... At present, we do not know the rules for which combinations are possible and used in coding the proteins."[15] The full explanation of how the information read from the genes is developed into parts and the whole of an organism should include [but does not yet] "the way in which the string of amino acids coded by the gene becomes a protein, that is, *a folded three-dimensional structure.* The sequence of amino acids is insufficient to explain this folding, and there are many alternative folded states for any sequence, only one of which is the physiologically active protein."[16] A gene should no longer be thought of as a linear string of DNA molecules, but as tri-dimensional. It is obvious that the tri-dimensional folding raises the complexity of these molecules enormously.

Taking all this into account, what remains of Dawkins' "replicators"? Dawkins had launched the term because the gene and its functions became more and more complex and mysterious; the term, in its abstraction, also proved useful to cover genes and memes at the same time. "The only kind of entity that has to exist in order for life to arise, anywhere in the universe, is the immortal replicator," and: "A gene is a replicator with a high copying-fidelity," he assured his readers.[17] But to replicate is an active verb, and Dawkins replicators were doubtlessly the direct offspring of his selfish genes, "the master programmers." Yet, it became ever clearer that the *genes do not play an active role* in the hereditary and life-building process. "It might be more useful to avoid saying that genes do anything at all; it is more that genes are used. They operate under control. ... Conditions in the cellular environment will switch a gene on or off to varying degrees." (Noble[18])

Genes *do* nothing: they are read by proteins called RNA which are under the control of the cell, which is under the control of the living organism in its environment. "Genes can *make* nothing," affirms Lewontin. "A protein is made by a complex system of chemical production involving other proteins, using the particular sequence of

15. Denis Noble: op. cit., p. 8.
16. Richard Lewontin: *The Triple Helix*, p. 115 (italics added).
17. Richard Dawkins: *The Selfish Gene*, pp. 266, 28.
18. Denis Noble: op. cit., p. 105.

nucleotides in a gene to determine the exact formula for the protein be-ing manufactured ... Nor are [genes] self-replicating. They cannot make themselves anymore than they can make a protein. Genes are made by a complex machinery of proteins that uses the genes as models for more genes. When we refer to genes as self-replicating, we endow them with a mysterious, autonomous power that seems to place them above the more ordinary materials of the body. Yet if anything in the world can be said to be self-replicating, it is not the gene, but the entire organism as a complex system."[19]

"There is no such thing as a self-replicating molecule in biology and probably there never has been," concurs Gabriel Dover.[20] As pointed out before, if the gene had any deciding power in its actions, as Dawkins selfish-gene clearly has, it would be a replacement of a Creator, De-signer, or Life Force, without any explanation of how this came to be. "Genes cannot be selfish or unselfish," writes Mary Midgley, "any more than atoms can be jealous, elephants abstract or biscuits teleological." Therefore the totality of an organism's genes, its genome, "is not under-standable as 'the book of life' until it is 'read' through its 'translation' into physiological function. My contention," states Denis Noble, "is that this functionality does not reside at the level of the genes. It can't because, strictly speaking, the genes are 'blind' to what they do ..."[21] "It takes more than DNA to make a living organism," writes Lewontin. "Even the organism does not compute itself from its DNA. A living organism at any moment in its life is the unique consequence of a devel-opmental history that results from the interaction of and determination by internal and external forces."[22]

As the gene, even in its disguise as a replicator, was being put in its place, other elements of life gained in importance: the cell, living element of all organisms, the organism itself, and the environment in which it lives. "The genes cannot do what they do without the proteins. And the proteins are not free agents, either. They respond to influences from across the rest of the organism and ultimately from the environment too." And then we read this crucial statement by the Emeritus Professor of Cardiovascular Physiology at the University of Oxford, Denis Noble: "After all, the genes by themselves are dead. It is only in the fertilized egg cell, with all the proteins, lipids, and other

---

19. Richard Lewontin: *The Doctrine of DNA*, p. 48 (italics in the text).
20. Hilary and Steven Rose: *Alas Poor Darwin*, p. 51.
21. Denis Noble: op. cit., p. 34.
22. Richard Lewontin: op. cit., p. 63.

cellular machinery *inherited from the mother*, that the process of reading the genome to initiate development can get going. ... The expression of a gene will involve levels of activity that are determined by *the system as a whole*. This is so obvious that it is truly extraordinary that there should be such great and repeated need to point this out."[23]

Life is transmitted by life. Although its origin remains unknown, life on Earth has been transmitted for about four billion years through living, self-replicating cells (not genes) which, at one point in the evolution, formed living, self-replicating organisms. The problem is not how tiny bits of matter could become alive, for they cannot; the problem is how life has used matter for its embodiment. "DNA never acts outside the context of a cell. And we inherit much more than our DNA. We inherit the egg cell from our mother with all its machinery, including mitochondria, ribosomes, and other cytoplasmic components, such as the proteins that enter the nucleus to initiate DNA transcription."[24]

"And yet," adds Noble, "the central biological dogma of our time [Crick's central dogma] is that inheritance is solely through DNA."[25] If this is wrong, what is then the role of the genes, of DNA, in the unfolding event of terrestrial evolution? As Stephen Gould put it: "The DNA does not determine a species, it is the record of a species." In other words: "Gene differences do not cause evolutionary changes in populations, they register those changes."[26] The genes contain the information of the fundamental plan, acquired through evolutionary heredity, on which a species is built, which is passed on from generation to generation often for five million years or more, and whose extreme complexity is read for the building and keeping alive of an organism that is still more complex, for it may consist of billions and even trillions of cells, in one way or another responding to each other and working together.

Lamarck's Come-Back

We have noted before that Lamarckism comes knocking at the door every time the conditions of our planet, which are the environment of its organisms, is paid attention to. Weismann and Crick put an insurmountable barrier between the organism and its environment. Only

---

23. Denis Noble: op. cit., pp. 45-6 (italics added).
24. id., p. 41.
25. ibid.
26. Kim Sterelny: *Dawkins vs. Gould*, p. 83.

the genes and their mutations were relevant to the evolution of the species and the survival of the fittest, in which the external surroundings played no part at all. Dawkins even expressed the bizarre opinion that the bodies of organisms did not have to exist, "the immortal replicators" would suffice for the job. One wonders how a world peopled by replicators might look like. How would they manage to write a book like *The Selfish Gene*?

The cutting edge of biological research at present seems to re-establish the obvious and necessary interaction between the organism and its environment. Adaptation and natural selection are clearly related to the living conditions (niche) of the species, in a permanent confrontation and exchange of the organisms with their surroundings. As important is the fact that a living organism is not a material object moved only by external forces, but a living thing that reacts to the external circumstances. The flow of the processes of life can no longer be seen in the reductionist way as going from the genes to the proteins, cells, tissues, organs and organism, but is at each step determined by the more encompassing order and therefore going in the inverse way. "The genes cannot do what they do without the proteins. And the proteins are not free agents either. They respond to influences from across the rest of the organism and ultimately from the environment too."[27]

'Darwinism' "alienated the inside from the outside, by making an absolute separation between the internal processes that generate the organism and the external processes, the environment, in which the organism must operate," writes Lewontin. He recognizes this reductionist procedure as "an absolutely essential step in the development of modern biology. Without it, we would still be wallowing in the mire of an obscurantist holism that merged the organic and the inorganic into an unanalyzable whole." But "the conditions that are necessary for progress at one stage in history become bars to further progress in another"[28] – at which we now seem to be arriving.

The interest in Jean-Baptiste de Lamarck's theory is growing. The principal point of his thesis was that an organism can acquire external characteristics, which become part of its constitution. The source of such an acquisition is an inner *need* – in French *besoin* – created by the circumstances (and not an inner 'will' as is so often supposed). This has been trenchantly and sarcastically opposed by the reductionist mind on

27. Denis Noble: op. cit., p. 105.
28. Richard Lewontin: *The Triple Helix*, pp. 42, 47.

the ground that organisms, which consist of matter, cannot have intentions, or a will, or a need. But then the same reductionist mind has been time and again contradicting itself when it accepted the need to feed, to reproduce and, as the fittest, to gain the upper hand in the evolutionary race. The misunderstandings about "the great taboo that is called Lamarckism" have been considered in the second chapter of this book. Joining André Pichot, whose historical research seems to be spreading also in Anglo-Saxon countries, Denis Noble writes: "Neither Darwin nor Lamarck would recognize this travesty of biological thought,"[29] which has been going on since Weismann, i.e. for more than a century.

"A mother transmits to the embryo adverse or favourable influences on its gene expression levels," writes Noble. "These influences, called 'maternal effects', can even extent over several generations. ... Inheritance of this kind forms no part of neo-Darwinian theory. On the contrary, it is close to the great taboo that is called 'Lamarckism'. ... A lot of effort is now being devoted to exploring such effects. We are at the beginning of what may be a long and exciting process of discovery. ... Lamarckian inheritance would not exclude Darwinian selection. It would complement it, providing yet another source of diversity. ... Moreover, the expression or repression of genes may be affected by experience in a previous generation."[30]

The "Holy Grail" of Biology

"By the late 1980s it was becoming obvious to most genetic researchers that the heroic effort to find the information specifying life's order in the genes had failed," writes Michael Denton. "There was no longer the slightest justification for believing that there exists anything in the genome remotely resembling a programme capable of specifying in detail all the complex order in the phenotype [the living organism]. ... It is true that genes influence every aspect of development, but influencing something is not the same as determining it."[31] Still, there seems to be a marked difference between the mentality and knowledge of the researchers in the frontline of biology, and the noise generated by some bestselling science authors and the media. NASA is expert at advertis-

29. Denis Noble: op. cit., p. 99.
30. id., pp. 48, 95, 19.
31. William Dembski (ed.): *Uncommon Dissent*, p. 172.

ing its enterprises, sometimes more with an eye to the funding of its hugely expensive organization and projects than to the information of the public. The human genome project was launched with similar ballyhoo, and announced as no less than "the Holy Grail" of the biological sciences.

A genome is the complete set of genes contained in the nucleus of each of the cells of an organism. It is "all the genes of an organism together." The DNA molecules of a gene are wound like a double helix consisting of four nucleotides arranged in pairs. In the microscopic nucleus of a cell there are an enormous numbers of such molecules, folded tri-dimensionally, and forming the chromosomes, of which each human body cell contains 46. As the orthodox view of genetics still holds that the formation of an organism is directly based on the genes, the location of their molecules and the sequences in which they are ordered is considered of prime importance.

A few years before the completion of the human genome "one of the most eminent molecular biologists, Sydney Brenner, speaking before a group of colleagues, claimed that, if he had the complete sequence of DNA of an organism and a large enough computer, he could compute the organism. ... A similar spirit motivates the claim by yet another major figure in molecular biology, Walter Gilbert, that, when we have the complete sequence of the human genome, 'we will know what it is to be human'."[32] Such were the claims and the expectations a decade ago. Richard Dawkins, not to be outdone, wrote: "I conjecture that an embryologist of 2050 will feed the genome of an unknown animal into a computer, and the computer will simulate an embryology that will cumulate in a full rendering of the adult animal. ... We shall feed the genome of an unknown animal into a computer that will reconstruct not only the form of the animal but the detailed world in which its ancestors lived, including their predators or prey, parasites or hosts, nesting sites, and even hopes and fears."[33]

The mapping and sequencing of the human genome, an international undertaking, was completed in 2003. Our genome is 3 billion base pairs long, which form 20 000 to 30 000 genes. The reason why the last number remains undetermined is precisely what we have met with in the section on the genes: there is no one-to-one effect, different genes are read simultaneously to form the proteins and their constructions,

---

32. Richard Lewontin: *The Doctrine of DNA*, pp. 10-11.
33. John Brockman (ed.): *The Next Fifty Years*, pp. 154-55.

and more and more new mechanisms are being discovered which influence the action of the genes or are necessary for their expression. It did not take long, after all the festivities and media hype around the completion of the sequencing of the human genome, for the volume of the claims to die down. Who remembers the words of President Nixon when he declared the first landing of a human being on the Moon as important as the creation of the universe? Now the Holy Grail of biology would give humankind the power over life (and the power over life means the power over death). Has it?

Michel Morange began his recent book *Les secrets du vivant* (The Secrets of Life) with the words: "Some years ago the programme of sequencing of the human genome was presented as if it would teach us everything about ourselves. It would be 'the Grail of human genetics', according to the American geneticist Walter Gilbert. Today, however, knowing the genome seems, in the words of another American biologist, David Baltimore, the starting point of the post-genomic studies which will reveal to us the foundations of life, and also give us the understanding of our human nature. How to explain such a rapid change?"[34]

Walter Gilbert, prominent promoter of the HGP, the human genome project, is quoted as saying that, when the genome would be known, one would be able to pull a CD out of one's pocket and say: "Here is a human being; it's me!" "Today," writes Fox Keller, "almost no one would make such a provocative claim. Doubts about the adequacy of sequence information for an understanding of biological function have become ubiquitous, even among molecular biologists, and largely as a consequence of the increasing sophistication of genomic research. ... To an increasingly large number of workers at the forefront of contemporary research, it seems evident that the primacy of the gene as the core explanatory concept of biological structure and function is more a feature of the twentieth century [i.e. the past] than it will be of the twenty-first."[35]

Epigenetics

"The notion of the gene has been considerably modified because of the progress of our knowledge, so much so that today everyone

---

34. Michel Morange: *Les secrets de la vie*, p. 5.
35. Evelyn Fox Keller: *The Century of the Gene*, pp. 14, 9.

understands it in his own way, and that there are more questions than answers," writes Claude Lafon. He adds: "It is thought that we know everything about genetics, and yet we are entering a new era of one new and crucial problem after the other."[36] According to Henry Bauer's description of the filtering down of a new idea in science, the signs of a new era are not to be sought for in the textbooks or in popular introductions, but in the science magazines. We are also aware that new eras are often announced but almost as often fail to materialize.

In an August 2002 issue of the French magazine *Sciences et Avenir* (science and future), we read the exclamation: "DNA is not all! Researchers are on the track of a second biological code, contained in the proteins which fashion the DNA into chromosomes: histones." Given the existing barriers and dogmas, a second biological code contained in the proteins is either a blunder or a genetic revolution. "It has been announced loud and clear: once the DNA molecule would be decoded [as happened in 1953], once its language would be interpreted and copied, the gates of the alchemy of life would be wide open to us. We would then know how an organism is built, how the cellular mechanism keeps it alive, and above all how it might be cured. Alas, our hopes remain unfulfilled – or at least somewhat premature. For, indeed, the processing by the genes does not rest uniquely on the DNA." (Hervé Ratel[37])

In its November 2003 issue *Science et Avenir* publishes a file on genetics titled: "Crisis in genetics – Decline of the DNA empire." The banner reads: "The 'molecule of life' [namely DNA] is fighting for its life. Not a week goes by without the announcement of other important factors in heredity ..." The list of new factors in recent years is indeed amazing: prions, introns, transposons, histones, messenger RNA, transfer RNA, ribosomal RNA, soluble RNA, RNA interference, mitochondrial DNA, and so on. "Today," says Pierre Sonigo, "it becomes evident that the regulations within the cell – and consequently within the organism – are anything but linear [i.e. one-to-one]. They are part of series of loops without end and inextricable metabolic bifurcations, so that it gets very difficult to determine who does what and controls whom. Anyway, considering the latest research, one conclusion becomes evident: we must leave the notion 'DNA is everything' behind us; if not, research in molecular biology may remain marching on the spot for a long time to come."[38]

---

36. Claude Lafon: *Idées reçues en biologie*, p. 69.
37. *Sciences et Avenir*, August 2002, p. 78.
38. id., November 2003, pp. 56, 60.

Interviewed for the composition of this file, André Pichot confirms that in the laboratories the genes are no longer considered the active, determining builders of the organism. "The DNA is not a program but a data bank where the contents of the cell find their necessary information." In the words of Denis Noble, the genes "are read by proteins called RNA which are under the control of the cell." This, in its turn, reminds us of Stephen Gould's formulation: "The DNA does not determine a species, it is the record of a species." In other words: "Gene differences do not cause evolutionary changes in populations, they register those changes."

Epigenetics, though recognizing the importance of the genes, focuses more and more on the cell as a whole and on the environment. The following definitions of this new branch of the biological sciences may give an idea. It is "a new mode of heredity no longer based on the DNA. ... The term 'epigenetics' is used to describe the exceptional phenomena which do not fit into classic genetics coded by the DNA. ... Epigenetics deals with how gene activity is regulated within a cell. ... The epigenetic modifications control the expression of the genetic information, and are partially under the influence of the environment." According to one expert "the trickle of findings of epigenetic inheritance" two or three decades ago "is turning into a flood."[39]

Science et Avenir kept its readers informed about the ongoing revolution in genetics by publishing a new file in March 2008: "Genes do not explain everything." "All is written in the DNA, all is contained in the DNA," says the introductory article. "This is the refrain that has been song to us since the structure of this molecule was decoded by James Watson and Francis Crick. ... Popularized to excess by the scientists, industrialists and media, the gene, yesterday a certainty, today does not mean much anymore. ... Focusing excessively on the DNA has quite simply made us forget that it was nothing but one molecule among the thousands contained in a cell. No doubt, it is an important one. But it is totally incapable to describe our character, our ways of behaving, and the changes which our organism is to undergo when confronted with the aggression of the environment in the course of the years."

This file, put together by a popular science magazine aware of its responsibilities, stresses time and again the revolutionary character of the new genetic discoveries. We read about "epigenetics gaining the upper hand," and "the revolution in biology." "What is happening at

---

39. *New Scientist*, 9 July 2008.

present is a readjustment of our vision, a kind of salutary stepping back towards the beginnings of genetics before it became synonymous with DNA." And we are informed that, in December 2007, the geneticist Anne Plessis has organized *Les Masterales*, a series of lectures under the significant title: "The end of the dogmas in molecular biology." In an interview, the prominent biophysicist Henri Atlan says: "We must get rid of the fetishism of the gene,"[40] which reminds us of Pichot's "DNA mysticism" and "DNA mania."

French science finds it easier to take its distance from Darwinism, authentic or pseudo. As the aforementioned references are mainly from French sources, they should be counterbalanced by Anglo-Saxon ones. The following are culled from the online edition the popular American magazine *New Scientist*. Its issue of 9 July 2008 carries the title: "Rewriting Darwin: The new non-genetic inheritance." It starts as follows: "Half a century before Charles Darwin published *On the Origin of Species*, the French naturalist Jean-Baptiste de Lamarck outlined his own theory of evolution. The cornerstone of this was the idea that characteristics acquired during an individual's lifetime can be passed on to their offspring. Lamarck's theory was ignored or lampooned. ... In recent years, ideas along the lines of Richard Dawkins's concept of the 'selfish gene' have come to dominate discussions about heritability, and with the exception of a brief surge of interest in the late 19th and early 20th centuries, Lamarckism has been consigned to the theory junkyard.

"Now all that is changing. No one is arguing that Lamarck got everything right, but over the past decade it has become increasingly clear that environmental factors, such as diet and stress, can have biological consequences that are transmitted to offspring without a single change to gene sequences taking place. However, fully accepting the idea, provocatively dubbed 'the new Lamarckism', would mean a radical rewrite of modern evolutionary theory. Not surprisingly, there are some who see that as heresy. 'It means the demise of the selfish-gene theory', says Eva Jablonka at Tel Aviv University, Israel. 'The whole discourse about heredity and evolution will change.'"[41]

"There is more than a little irony in the present state of affairs," writes Evelyn Fox Keller, "for never in the history of the gene has the term had more prominence, in both the scientific and the popular press. ... For the basic fact is that, at the very moment in which gene-talk has

---

40. *Sciences et Avenir*, March 2008, p. 63.
41. *New Scientist*: op. cit.

come to so powerfully dominate our biological discourse, the prowess of new analytic techniques in molecular biology and the sheer weight of the findings they have enabled have brought the concept of the gene to the verge of collapse. ... We find that the gene has become many things – no longer a single entity but a word with great plasticity, defined only by the specific experimental context in which it is used."[42]

Of the ongoing revolution in biology, the general public is not yet informed. According to Kuhn's theory of the scientific paradigms, and the instances from the history of science, the resistance against a fundamental new idea is as ruthless as a fight for life or death. The media are among the most conservative carriers of information where science is concerned, for they must make sure to have it right and are wary of being tricked by wild new ideas. The progress of "epigenetics" will be interesting to follow. But the new scientific data, "a trickle that has become a flood," are too many and too well-founded to be withheld consideration. What is happening is that the strict, dogmatic tenets of positivism, reductionism, and scientific materialism as whole, are questioned because they no longer agree with the facts. This kind of revolution happened in physics a century ago. Biology may finally catch up, and plenty of surprises are awaiting us in this field.

---

42. Evelyn Fox Keller: op. cit., pp. 68-9.

13.

Science and Religion

After all, who or what besides myself will decide what I accept as truth? I know that if there are ultimate answers I'm surely not the court of last appeal as to what the answers are, but here in my study, on this human level, for myself, it appears I am. Does it matter very much what I decide? Not to science. But religion would have me think that the decisions of this private court when it comes to whether or not I will believe in God are of inestimably great significance.

KITTY FERGUSON

Any type of dogmatism is the very antithesis of science.

VICTOR STENGER

The Western God

"I am attacking God, all gods, anything and everything supernatural, wherever and whenever they have been or will be invented." It is remarkable that these passionate words of Richard Dawkins, already quoted previously, are from a man who considers himself a scientist, and whose name is now often conjoined with the name of Charles Darwin. "I have never heard such hardline, aggressive promotion of atheism under the guise of science as I have heard from the Darwinists," writes Denyse O'Leary. "It is, at best, amusing to hear Darwinists charge that the creationists have an underlying religious agenda, when the Darwinists own anti-religious agenda is pretty obvious."[1]

The literature produced by zealous atheist scientists, mainly biologists, is increasing steadfastly. Some telling titles are: *A Devil's Chaplain* and *The God Delusion* (Richard Dawkins, 2003 and 2006), *Breaking the*

---

1. Denyse O'Leary: *By Design or by Chance?* p. 239.

Spell (Daniel Dennett, 2006), *The End of Faith* (Sam Harris, 2006), *God is not Great* (Christopher Hitchens, 2007), *God – The Failed Hypothesis* (Victor Stenger, 2008). In all these books religion is presented as irrational, by which they are continuing the tradition of the Enlightenment; totalitarian, reigning by dogma and fear as well for the earthly as the eternal destiny; and outright evil, given the uncountable instances of physical and moral cruelty in history and even today. "Religion is an insult to human dignity. With or without it, you'd have good people doing good things and evil people doing evil things. But for good people to do evil things, it takes religion." (Steven Weinberg[2])

It is noteworthy that the controversy between religion on the one side and atheism and anti-theism on the other was and is taking place almost exclusively in the West, within the culture dominated by Christianity and having its roots in Europe. Modern culture, carrier of Christianity, the Enlightenment values and science, has spread over the globe from Europe, which deemed itself superior to all the peoples and cultures it discovered, conquered and exploited during the era of colonialism. The general outlook was Eurocentric, an attitude which has broadened into 'Western' when America entered the global stage, and which is still alive, consciously or subconsciously, as proved by the ongoing attack of science on religion.

The God in the literature mentioned above is invariably the bearded Man in nightdress Upstairs, a kind of irresponsible autocrat. "A naïve Western view of God is an outsize, light-skinned male with a long white beard, who sits on a very large throne in the sky and tallies the fall of every sparrow." (Carl Sagan[3]) In his Gifford lectures, the theologian Etienne Gilson argued: "Christian thought did not simply cloak itself in Greek philosophical ideas. What it uniquely added to Western philosophy was the Hebrew Creator."[4]

Like so many other concepts in the Christian religion, the mental idea of 'God' is, in itself, confused and contradictory. Arthur Lovejoy called "the word 'God' in the last degree ambiguous." In the Bible we find at least three different kinds of God: the tribal terrible, jealous and vengeful Yahweh, the metaphysical God of the prophets, and the loving God, the Father, of Jesus Christ. To these quite different Gods the Church Fathers added the Trinity, the three-in-one God.

The tribal God belonged to a period of the establishment of morality

---

2. Richard Dawkins: *The God Delusion*, p. 249.
3. Carl Sagan: *The Varieties of Scientific Experience*, p. 149.
4. Larry Witham: *The Measure of God*, p. 203.

among the primitive Hebrews, the new moral precepts being hewn in stone as the Ten Commandments. It was the tribal God who ordered: "If your brother, the son of your father or of your mother, or your son or daughter, or the spouse whom you embrace, or your most intimate friend, tries to secretly seduce you, saying: 'Let us go and serve other gods', unknown to you or your ancestors before you, gods of the peoples surrounding you, whether near you or far away, anywhere throughout the world, you must not consent, you must not listen to him; you must show him no pity; you must not spare him or conceal his guilt. No, you must kill him, your hand must strike the first blow in putting him to death and the hands of the rest of the people following. You must stone him to death, since he has tried to divert you from Yahweh your God." This passage from the book *Deuteronomy* is here quoted from Sam Harris' *The End of Faith*. Lots of passages from *Deuteronomy* and other books constituting the Bible are ready ammunition for use by the anti-religious, not entirely without justification, for the Bible has provided, in the West, the religious and moral inspiration for centuries.

Another easily accessible provision of anti-religious ammunition is the history of Christianity. As soon as Catholicism was recognized as the official church of the Roman Empire, around the year 400, hordes of 'monks' went on the rampage, murdering 'heathens' and destroying the buildings of the existing religions and cults. There has been the Crusades, the Inquisition (still extant), the witch hunts which killed hundreds of innocent women in the cruellest manner, and the burning of unbelievers from within and without the Church's ranks. Dogmatic fanaticism brought Galileo to trial, "the most tragic event in the whole of the scientific revolution." Less well-known is the anti-modernist action of the Catholic Church following the declaration of papal infallibility in 1870. "The Vatican did not merely now make the claim of papal infallibility. In the nineteenth century, in its assaults on every development in scientific knowledge, every glimmering of light shed in the field of biblical scholarship, every advancement of technical skill (it even issued condemnation of the electric light), the Vatican was the great powerhouse of reaction, posing very grave difficulties for those who wished to practice the Catholic faith without committing intellectual suicide." (A.N. Wilson[5])

The anti-religious attitude of many scientists has been sharpened by the 20th century trials about the teaching of evolution in the

---

5. A.N. Wilson: *God's Funeral*, p. 286.

educational institutions of the USA. Christians tried to have the teaching of evolution forbidden by law or demanded equal time for biblical creationism. The most famous trial remains the State of Tennessee vs. John Scopes in 1925, known as "the monkey trial." In recent years, against the background of the controversy between creationism and "intelligent design," there have been several more court cases in which evolution was at stake. Such events seem to be spreading to other countries, sometimes instigated by Muslim students or teachers, for Islam does not accept the theory of evolution.

Besides, what doubtlessly intensified the motivation of the anti-religious was the spectacular destruction by Muslim terrorists of the twin towers of the World Trade Centre in New York on 11 September 2001. This blatant act of fanatical religious aggression made the relation of the Western mentality, tempered by the Enlightenment, and the dogmatism of the Islamic faith, professed by millions in several countries, into a red hot world issue. "9/11" changed drastically the Western attitude towards Islam, whose millions were formerly supposed to exist behind an invisible wall in the stagnant confined universe of their faith. Suddenly the Muslim religion was studied, and Muslim immigrants in non-Muslim countries obtained equal rights with the local populations. The friction between Islam and the Western 'Satan' is far from ended, and hardly a day passes without the news bulletins carrying an item about head scarves, burqa dresses, mosques – not to speak about atrocious terrorist acts in most of which Muslims kill their brothers and sisters in the Faith, and of course the wars and lesser military operations, still ongoing.

A little knowledge of the Christian past may cause profound amazement about the similarities between the Muslim world as it still is today, and the Catholic medieval world surviving in so many aspects of the Western way of life, tempered by the ideals of the Enlightenment. Omnipresent was the Christian morality taught by the Holy Book, inspired by God. "All those books which the Church regards as sacred and canonical were written with all their parts under the inspiration of the Holy Spirit. Now, far from admitting the coexistence of error, Divine inspiration by itself excludes all error, and that also of necessity, since God, the Supreme Truth, must be incapable of teaching error." (Leo XIII, 1893[6]) Galileo barely escaped burning, Bruno did not. Descartes hesitated to publish, Newton hid his Arianism, Darwin stalled writing

---

6. Sam Harris: *The End of Faith*, p. 104.

what would become the *Origin* for twenty years, the theologians Küng and Schillebeeckx were forbidden to teach, liberation theology was forbidden, many of the most important writings were put on Rome's "Index of Forbidden Books." All this because of the Christian mullahs, madrasas, communal congregations, and fatwas.

The three Abrahamic religions – Judaism, Christianity and Islam – are highly esteemed by theologians because they are supposed to be the only monotheisms. This is a gross misconception. "The earliest Greek natural theology was certainly monotheistic ... The Greeks did not invent polytheism but instead spoke of a single god whom they call Zeus, whose mind embraces all things in its knowledge, and who guides all things and is king of all'." (Werner Jaeger[7]) And there was the Platonic Absolute, the God identified with the Idea of the Good as perfection or self-sufficiency. Moreover, one must be ignorant of Hinduism not to know about Brahman, the Absolute, of whom the gods and goddesses are the cosmic powers, and who is represented above the entrance of the temple of every Hindu god and goddess by the glyph for OM.

What the three Abrahamic monotheisms certainly have in common is their totalitarianism and the ruthlessness, often culminating in cruelty, of its enforcement. It may be remembered that the Hebrews were the Chosen People to whom all "the nations" were to bow, and whose reign over the Earth would be the sign for the Last Judgment. Nor should it be forgotten that Judaism had its own Galileo in the person of Baruch Spinoza (1632-77), one of the great philosophers. "On 27 July 1656, the elders of the Amsterdam synagogue made the following *cherem*, or damnation, or fatwa, concerning his work: 'With the judgment of the angels and of the saints we excommunicate, cut off, curse, and anathemize Baruch de Espinosa, with the consent of the elders and of all this holy congregation, in the presence of the holy books: by the 613 percepts which are written therein, with the anathema wherewith Joshua cursed Jericho, with the curse which Elisha laid upon the children, and with all the curses which are written within the law.

"Cursed be he by day and cursed by night. Cursed be he in sleeping and cursed be he in waking, cursed in going out and cursed in coming in. The Lord shall not pardon him, the wrath and fury of the Lord shall henceforth be kindled against this man, and shall lay upon him all the curses which are written in the book of the law. The Lord shall destroy his name under the sun, and cut him off for his undoing from all the

---

7. Larry Witham: op. cit., p. 41.

tribes of Israel, with all the curses of the firmament which are written in the book of the law.'"[8] Excommunication was common practice in the Catholic Church also. Some popes excommunicated whole nations, thereby, at least in intention, sending the souls of their people to burn in hell for eternity, and this every so often for the basest political or financial motives.

Where the anti-religious scientists go seriously wrong, however, is when they consider the mania shown by the religions to be an exclusively religious phenomenon instead of a way the human mind functions. Nowhere in the books mentioned at the beginning of this chapter does one read about the philosophical and political manias of Russian communism, Nazism or Maoism, or of the general aggressive egoistic attitude of individual versus individual, community versus community, caste versus caste, nation versus nation, culture versus culture. The egocentric, sectarian functioning of the mind is part of our evolutionary condition, as is its fear and insecurity. Education, inculcation and brainwashing construct a human mind in a way which is near to impossible to change, except by a long and painful conversion process. We have seen that conversion also plays a part in Thomas Kuhn's theory of the paradigms in science. And the attitude of the atheist and antitheist scientists becomes quasi farcical when they do not seem to realize that they themselves are defending a dogmatic totalitarianism that has become the Church of Scientism (not to be confused with Scientology). "It is easy to forget that both science and religion are preoccupied with justifying beliefs," (Steve Fuller[9]) which is putting it mildly.

Religion or Spirituality

"There is nothing accidental about the difference between a Church and its Founder," wrote Bertrand Russell. "As soon as absolute truth is supposed to be contained in the sayings of a certain man, there is a body of experts to interpret his sayings, and these experts infallibly acquire power, since they hold the key to truth. Like any privileged caste they use their powers to their own advantage."[10] This sequence of events has been repeated throughout history and *is* actually the history

---

8. Christopher Hitchens: *God is not Great*, pp. 314-15.
9. Steve Fuller: *Kuhn vs. Popper*, p. 16.
10. *Russell on Religion*, p. 10.

of religion: a Person has a new revelation or realization which attracts a group of followers; they lower his inspired message to their level of understanding and fix it into a set of formulas comprehensible by all; the new message is declared the absolute truth; this truth is to be approached solely through a body of initiated representatives, who invent a set of rules and rituals to be executed obligatorily. To this has to be added that an individual approach to the Truth is forbidden. Such, by and large, has also been the history of Christianity.

"The religious life is a movement of the same ignorant human consciousness, turning or trying to turn away from the earth towards the Divine, but as yet without knowledge and led by the dogmatic tenets and rules of some sect or creed which claims to have found the way out of the bounds of the earth-consciousness into some beatific Beyond," wrote Sri Aurobindo. "The religious life may be the first approach to the spiritual, but very often it is only a turning about in a round of rites, ceremonies and practices or set ideas and forms without any issue. The spiritual life, on the contrary, proceeds directly by a change of the consciousness, a change from the ordinary consciousness in which one finds one's true being and comes first into direct and living contact and then into union with the Divine."[11]

The spiritual life is an individual, inner exploration. The spiritual aspirant undertakes the adventure of the discovery of Reality by a direct contact, for the Real is present in himself. To start on his quest, he may follow the inspiration of persons who have preceded him on the different levels of the human personality. His first task is the mastery of the way evolution has produced our extremely complex being. This effort itself is already so daunting, especially in the nether vital and material regions, that several spiritual paths have limited their exploration to the more accessible regions of the soul and the mind.

Practitioners of the individual inner exploration are not welcome in the body of the established religious communities, ruled by dogma and authority. In the West, where they are called 'mystics', hardly a single one has escaped persecution by the Church, and many have paid with their life for the truthful confession of their spiritual attainments.

For scientific materialism there is no inner reality, and consequently no spirituality. Those who still believe so are retarded, mystics or charlatans. They are derided on practically every other page written by the positivist reductionists, who do not mind to flaunt their ignorance of

---

11. Sri Aurobindo: *Letters on Yoga*, p. 137.

such foolish if not cretinous matters. Take for instance Daniel Dennett, the American philosopher in the Dawkins camp. He concedes: "I simply do not know enough about religions to write with any confidence about them," but the attacks them nevertheless with candid confidence. "Perhaps," he writes, "I should have devoted several years more to study before writing this book," namely *Breaking the Spell* – the spell of religion. "I will try to tell the best current version of the story science can tell about how religions have become what they are. I am not at all claiming that this is what science has already established about religion. The main point of this book is to insist that we don't yet know the answers to these important questions. ...The spell that must be broken is the taboo against a forthright, scientific, no-holds-barred investigation of religion as one natural phenomenon among many."[12]

Besides, why should religion or spirituality remain of interest at all? "We scientists have the drama, the plot, the icons, the spectacles, the 'miracles', the magnificence, and even the special effects. We inspire awe. We evoke wonder. And we don't have one god, we have many. We find gods in the nucleus of every atom, in the structure of space-time, in the counter-intuitive mechanisms of electromagnetism. What richness! What consummate beauty! (Carolyn C. Porco[13]) The leading promoter of the poetry of science and poetry *in* science was Carl Sagan, who said: "Science arouses a soaring sense of wonder," and let no occasion pass by without trying to prove his point. He has been arduously imitated by the militant anti-religious camp, e.g. Peter Atkins, who exclaimed in an interview: "The world around us is extraordinary: it's delightful, it's wonderful, it's awesome. Science enables us to pick it apart, to look inside and to see why it is so wonderful."[14] In the works of thus-minded authors this praise of materialistic science is expressed again and again – till one reaches the pages on theodicy, where the horror of life and human existence are being discussed, and the incompetence of the Good God to change or put an end to them.

---

12. Daniel Dennett: *Breaking the Spell*, pp. 334, 103, 17.
13. John Brockman (ed.): *What is Your Dangerous Idea?* p. 156.
14. Russell Stannard: *Science and Wonders*, p. 166.

Empiricism and Religion

As we have seen, Dennett wants "to investigate religious phenomena scientifically" without knowing what they are. He praises "those pioneers who are now beginning for the first time really to study the natural phenomena of religion through the eyes of contemporary science." He wants "to put religion on the examination table."[15] Victor Stenger is of a similar mind: "If a person undergoes a religious experience that truly places him in communication with some reality from beyond the material world, then we may reasonably expect that person to have gained some deep, new knowledge about the world that can be checked against empirical facts."[16] However, the problem, which Stenger here curiously overlooks, is that "some reality beyond the material world" and the reality of "empirical facts" are in the eyes of scientific materialism two utterly different realms, of which the former is *a priori* censured by the latter.

Empiricism recognizes what it calls 'objective' facts perceived through the senses. Sri Aurobindo notes that "science cannot dictate its conclusions to metaphysics any more than metaphysics can impose its conclusions on science."[17] That matter is the only reality, is also the basic principle of the new field of 'neurotheology', in which scientists seek the biological basis of spirituality, asking questions like: Is God all in our heads? Countless articles in science magazines carry titles of the same sort: "How God lives in my right brain ... Do we have neurons specialized in the divine? ... The biology of religious faith ... Three religious experiences under the microscope ... Is fanaticism a molecule? ..." In France, Patrick Jean-Baptiste has even published a book with the title *La biologie de Dieu* – The Biology of God (2003). John Horgan, a journalist and contributor to *Scientific American*, wrote three books based on interviews with key figures in science and trying to fathom the heart of the matter. In the last one, *Rational Mysticism*, he comes to the conclusion: "Scientists studying mysticism are still in the fact-accumulation stage, and may always be. ... The fact is, neuroscientists cannot explain how the brain carries out the most elementary acts of cognition."[18]

Stephen Gould, an agnostic materialist, but a cultured one (which is

---

15. Daniel Dennett: op. cit., pp. 28, 31, 39.
16. Victor Stenger: *God – the Failed Hypothesis*, p. 171.
17. Sri Aurobindo: *The Life Divine*, p. 178 (footnote).
18. John Horgan: *Rational Mysticism*, pp. 136-37.

not as common as one might wish), has introduced a much discussed concept in his *Rock of Ages*: NOMA. "To say for all my colleagues and for the umpteenth millionth time: science simply cannot, by its legitimate methods, adjudicate the issue of God's possible superintendence of nature. We neither affirm nor deny it; we simply cannot comment on it as scientists."[19] NOMA stands for "non-overlapping magisteria," which means that science and religion both rule in their own domain and cannot or should not be in conflict. As Gould sees it – and many agree with him – the magisterium of science covers the empirical realm of fact and theory; the magisterium of religion extends over questions of ultimate meaning and moral value. (The term 'magisterium' is best understood as sphere of authority or domain of competence.)

"I do not see how science and religion could be unified, or even synthesized, under any common scheme of explanation or analysis," wrote Gould, "but I also do not understand why the two enterprises should experience any conflict. Science tries to document the factual character of the natural world, and to develop theories that coordinate and explain these facts. Religion, on the other hand, operates in the equally important, but utterly different, realm of human purposes, meanings and values – subjects that the factual domain of science might illuminate, but can never resolve. ... These two magisteria do not overlap ... To cite the old clichés, science gets the age of rocks, and religion the rock of ages; science studies how the heavens go, religion how to go to heaven."[20]

Knowing what we do about sociobiology, it is clear that its representatives could not possibly accept such a standpoint. According to them, biology, based on physics, is the science that will provide humanity with a total explanation and interpretation of what it is, where it is, and wherefore it is there. Such was the vision of Edward Wilson, and such is the still more extreme gospel of Richard Dawkins: "Science shares with religion the claim that it answers deep questions about origins, the nature of life, and the cosmos. But there the resemblance ends. Scientific beliefs are supported by evidence, and they get results. Myths and faiths are not and do not."[21]

Might there not be a way to bring science and religion to a common understanding? "Convergence? Only when it suits," answers Dawkins. "To an honest judge [which must be a scientific materialist, of course],

---

19. Richard Dawkins: *The God Delusion*, p. 55.
20. Stephen Jay Gould: *Rock of Ages*, pp. 4, 6.
21. Richard Dawkins: *River out of Eden*, p. 37.

the alleged convergence between religion and science is a shallow, empty, hollow, spin-doctered sham."[22] "Science and religion cannot be reconciled," concurs Peter Atkins, "and humanity should begin to appreciate the power of its progeny [science, that is] and refuse all efforts at a compromise. Religion has failed, and its failure should be brought into the open. Science, with its present successful effort at universal competence through identification of the smallest [reductionism], she who is the highest joy of the intellect, should be recognized as the universal Queen."[23] Steven Weinberg, caustically, is "all in favour of a dialogue between science and religion, but not a constructive dialogue. One of the great achievements of science has been, if not to make it impossible for intelligent people to be religious, then at least to make it possible for them not to be religious."[24]

The End of Science?

It is not the intention of this book to attack science, but to examine its claims to exclusivity and absolute truth. The question which is the title of this section may shock at first sight in a culture which prides itself on being scientific, but it is on the order of the day especially among theoretical physicists. The following recent book titles speak for themselves: *The End of Science* (John Horgan), *La fin des certitudes* (The End of Certainties, Ilya Prigogine), *Science in the Age of Uncertainty* (John Brockman, ed.), *The Trouble with Science* (Robin Dunbar), *The Trouble with Physics* (Lee Smolin), *The End of Physics: The Myth of a Unified Theory* (David Linley) ... Jean-Pierre Vigier, a French physicist, said that "physics is in crisis, we are in full struggle, the stakes are enormous." Steven Weinberg, Nobel Prize in physics, conceded that "it is a terrible time for particle physics."

Not only particle physics is in crisis. We find alarming words of desperation from prominent scientists in other fields which are commonly regarded as definitively acquired and established knowledge. "Perhaps the resolution [of some problems] is that our theory of gravity – the general theory of relativity – is just plain wrong." (David Susskind). "The results [of some experiments] could send us back to the drawing

---

22. Richard Dawkins: *A Devil's Chaplain*, p. 179.
23. John Lennon: *Hat die Wissenschaft Gott begraben?* p. 7.
24. John Horgan: op. cit., p. 86.

board about the early universe." The Big Bang, multiple universes, black matter and energy in macrophysics; the unification of relativity and the quantum theories; string theory, the search for the Higgs boson and the problem of reality in microphysics – these are only some of the problems now in need of an urgent solution.

"Science is a human institution, subject to human foibles – and fragile, because it depends as much on group ethics as on individual ethics. It can break down, and I believe that it is doing so now. ... There can be no doubt that we are in a revolutionary period. We are horribly stuck, and we need real seers, and badly," writes Lee Smolin, one of the founders of string theory who has become its severe critic. And he reminds us that "there is no scientist, not even Newton or Einstein, who was not wrong on a substantial number of issues they had strong views about. ... There is a great tendency to think that the fundamental principles of physics, once discovered, are eternal, yet history tells us a different story. Almost every principle once proclaimed has been superseded."[25] Nowadays the influence of Thomas Kuhn's theory of the paradigms, the changing views of the basics of science, is felt everywhere.

As this issue is so important, and for the most part still unrealized by the educated public, we quote the statements of two other scientists. The first one is Ilya Prigogine, Nobel Prize winner. "At the macroscopic and microscopic levels, the natural sciences have rid themselves of a conception of objective reality that implied that novelty and diversity had to be denied in the name of immutable universal laws. They have rid themselves of a fascination with a rationality taken as closed and a knowledge seen as nearly achieved. They are now open to the unexpected, which they no longer define as the result of imperfect knowledge or insufficient control."[26] In another essay he writes: "The fundamental laws are now the expression of possibilities and no longer certitudes ... I think that we are only at the beginning of the adventure. We are witnessing the emergence of a science which is no longer limited to simplified, idealized situations, but which puts us face to face with the complexity of the real world ..."[27]

Another voice is that of Sven Ortoli, a journalist trained as a physicist, who says: "Determinism is badly shaken. What is more, there are some scientists, Nobel Prize winners among them, who go far enough to consider the universe as an essentially spiritual phantasmagoria. The

---

25. Lee Smolin: *The Trouble with Physics*, pp. 308, 311, 275, 218.
26. Ilya Prigogine and Isabelle Stengers: *Order out of Chaos*, p. 306.
27. Ilya Prigogine: *La Fin des certitudes*, pp. 14, 16.

majority of the physicians reject such an extreme hypothesis, but this does not prevent it from being there and from being accepted by what may still be considered demented beliefs, to the outrage of the defenders of intellectual traditions dating back to the 19th century. ... The physics of the last century has been completely destroyed ... The facts that those pieces of matter [the elementary particles] have proved to be in reality nothing but mathematical abstractions, non-local, which means that they can spread out over the whole of space, and that they obey determinism no longer, has given a fatal blow to 'classic' materialism. True, materialism is still possible, but then 'quantum' materialism, which should be called 'fantastic materialism' or 'science fiction materialism'." And he goes still further: "The idealism that believes in the autonomous existence of the spirit, comes to the fore again. A kind of new religion, which we have called 'quantum syncretism', is being born; it refers everything – matter and spirit – to an Absolute that is unknowable, but whose existence could be deduced from the extraordinary aspects of the new physics."[28]

Towards the end of his life, Albert Einstein once said: "You probably think that I look back on the work of my life with quiet satisfaction. Seen from nearby, it is totally different. There is not a single concept of which I am convinced that it will remain unchanged, and I feel uncertain to be even on the right path. ... If there is something that I have learned from the intricacies of a long life, it is that we are much farther from a more profound insight in the elementary complexities than most of our contemporaries think we are."[29]

But the main focus of our interest is on evolution. The question becomes more pressing: if the foundation of biology is physics, and physics seems badly shaken by the uncertainty about its underpinnings, how does biology react to this? The answer, as we have seen in Dawkins and his supporters, is quite simple: it takes no notice and looks the other way. His words may be recalled: "I am a biologist. I take the facts of physics for granted. If physicists still don't agree over whether those simple facts are yet understood, that is not my problem." If the accepted fundamental basis of biology is shaking, is that not his problem too, especially he being a reductionist?

Other biologists are more reasonable, or more sincere. "We know better than we did what we do not know and have not grasped. We

---

28. Sven Ortoli and Jean-Pierre Pharabod: *Le cantique des quantiques*, pp. 5, 6, 125.
29. John Lennon: op. cit., pp. 31, 32.

do not know how the universe began. We do not know why it is there. Charles Darwin talked speculatively of life emerging from 'a warm little pond'. The pond is gone. We have little idea how life emerged, and cannot with assurance say that it did. We cannot reconcile our understanding of the human mind with any trivial theory about the manner in which the brain functions. Beyond the trivial, we have no other theories. We can say nothing of interest about the human soul. We do not know what impels us to right conduct or where the form of the Good is found." (David Berlinski[30])

Scientists Pro and Contra

Ken Wilber was one of the thinkers who reacted strongly against the thesis of popular books like *The Tao of Physics* and *The Dancing Wu Li Masters*. He "disagreed entirely" with such books "which had claimed that modern physics supported or even proved Eastern mysticism. This is a colossal error. Physics is a limited, finite, relative, and partial endeavour, dealing with a very limited aspect of reality. It does not, for example, deal with biological, psychological, economic, literary, or historical truths; whereas mysticism deals with all that, with the Whole."[31]

One of Wilber's least known books is *Quantum Questions*, in which he examines the sources of the thought that created 20th century physics: Einstein, Eddington, Bohr, Heisenberg, Schrödinger, Pauli ... some of whom he quotes extensively. "Everyone of the physicists in this volume was a mystic," he writes. "They simply believed, to a man, that if modern physics no longer objects to a religious worldview, it offers no positive support either; properly speaking, it is different to all that." It is different because it was in the past and is still at present a work of the mind, and the mind is only part of the Whole, incapable of grasping the Whole. "They all shared a profoundly spiritual or mystical worldview, which is perhaps the last thing one would expect from pioneering scientists."[32]

Wilber quotes Arthur Eddington: "Briefly the position is this. We have learned that the exploration of the external world by the methods of physical science leads not to concrete reality but to a shadow world

---

30. David Berlinski: *The Devil's Delusion*, p. xiii.
31. *The Essential Ken Wilber*, p. 19.
32. id., p. 16.

of symbols, beneath which those methods are unadapted for penetrating. Feeling that there must be more behind, we return to our starting point *in human consciousness*, the one centre where more might become known."[33] A mathematical formula can never tell us what a thing is, only how it is moved. It can only specify an object through its external properties: movement, measure, mass.

Wilber also quotes Erwin Schrödinger: "The scientific picture of the real world around me is very deficient. It gives a lot of factual information, puts all our experience in a magnificently consistent order, but it is ghastly silent about all and sundry that is really near to our heart, that really matters to us. It cannot tell us a word about red and blue, bitter and sweet, physical pain and physical delight; it knows nothing of beautiful and ugly, good or bad, God and eternity. Science sometimes pretends to answer questions in these domains, but the answers are very often so silly that we are not inclined to take them seriously. ... In brief, we do not belong to this material world that science constructs for us. ... From where do I come and to where do I go? That is the great unfathomable question, the same for everyone of us. Science has no answer to it."[34]

This is of course a tone which differs altogether from much that we have heard before. These physicists, among the very greatest, dared to reflect and to speak out on the essential problems of our lives and on the relation of science to them. The difference between them and the parochial reductionists we have become acquainted with, is considerable. Moreover, "their writings are positively loaded with references to the Vedas, the Upanishads, Taoism (Bohr made the yin-yang symbol part of his family crest), Buddhism, Pythagoras, Plato, Plotinus, Berkeley, Schopenhauer, Hegel, Kant, virtually the entire pantheon of perennial philosophers."[35]

Consequently Wilber divides the 20th century physicists into two batches: the open-minded "mystics" including all those named above, and the mathematical theorists like Dirac, Weinberg, Feynman and Witte, plus most of the physicists active in the near past and at present. As to the latter, one quote from Steven Weinberg says it all: "Among today's scientists I am probably somewhat atypical in caring about such things [the concept of God]. On the rare occasions when conversations over lunch or tea touch on matters of religion, the strongest reactions

---

33. Ken Wilber: *Quantum Questions*, p. 10 (emphasis added).
34. Ken Wilber: op. cit., pp. 81, 83.
35. id. p. 6.

expressed by most of my fellow physicists is a mild surprise and amusement that anyone still takes all that seriously. ... As far as I can tell from my own observations, most physicists today are not sufficiently interested in religion even to qualify as practicing atheists."[36]

Whence this huge existential and perspectival difference? The "mystic" physicists lived on the fault line between two eras, between two Kuhnian paradigms in science. They personified the transition between the Newtonian era, as it were solidified by the 19th century positivism we have met in the lives of Lamarck, Darwin and Wallace, and their own 20th century thinking which put everything into question. An important factor here is that religion was no longer part of the equation. Spirituality, or "mysticism", or "the oceanic feeling," yes; dogmatic religion, no. For Einstein, Eddington, Heisenberg, Schrödinger, Pauli – to name only the best-known – re-thinking the universe in the terms of physics was their life task. In their quest, time and again, they found resonances and references in the testimonies left behind by others who had undertaken a similarly daunting task, though necessarily in other terms: the "mystics" in West and East, who had put *their* life on the line for similar reasons. Those who say that physics (and science in general) has nothing to do with "mysticism" are ignorant of its copiously documented history of a century, and less than a century, ago.

It may, moreover, be remembered that science was born within religion and in reaction to it. Galileo was as good a Catholic as any, for instance as the late-medieval monks who, together with the great Greeks, were his predecessors; but he shared the Renaissance mind to ask questions and to dare to answer them, even if they did not agree with the teachings of Rome. Johann Kepler, the great astronomer and mathematician so often ignored, was a mystic pure and simple, and spoke out as such, for instance in his *De Harmonia Mundi*. Descartes had been educated by the Jesuits, spent much of his life in hiding for fear of suffering the same treatment by the Inquisition as Galileo, and did his best to find room for God in his worldview. Newton, as is now well known, dedicated more years of his life to alchemy and biblical theories than he did to science, and all that *after* the publication of the *Principia Mathematica*. The erosion of God in the Western mind was the work of the Enlightenment, followed up in this by 19th century positivism, not by the actual founders of science, who would have been amazed at the consequences of their genius.

---

36. Steven Weinberg: *Dreams of a Final Theory*, p. 205.

Fundamentally, science is the search for Truth through the instrument of the mind. Religion and spirituality are the search and the effort to know and to live Truth through the instrument of the whole being, including the mind. Therefore there should be no contradiction between science, religion and spirituality. The attitude of the one or the other becomes twisted when one or the other claims an absolute prerogative, as do many religions in their Churches and as do the sciences in the Church of Scientism. This selfish, egocentric attitude is common to everything naturally human, because it is common to life's evolution of which we are the children.

The human being is human in the measure that it can overcome its selfishness – curiously enough the argument of Richard Dawkins when he divides the human personality in the part that is subject to the selfishness of the genes and the part of the memes that belongs to the other, cultural half. As Thomas Kuhn has made clear in *The Structure of Scientific Revolutions*: no scientific truth is absolute; all theories and paradigms are a partial approach. As Sri Aurobindo has made clear on the very first page of his *Essays on the Gita*: no written truth or holy book contains the absolute spiritual truth. Truth, to be known, has to be realized, and as such is always an approach, conditioned by the earthly circumstances of the beings who dedicate their life to this kind of realization.

There are even now many scientists who are religious. One is the astronomer Owen Gingerich, who writes: "As a scientist I accept methodological naturalism as a research strategy. ... There is no contradiction between holding a staunch belief in supernatural design and working as a creative scientist ... No one illustrates this point better than the seventeenth century astronomer Johann Kepler ... Kepler's work and life provide central evidence that an individual can be a creative scientist and a believer in divine design in the universe, and that indeed the very motivation for the scientific research can stem from a desire to trace God's handiwork. ... I think my belief makes me no less a scientist. ... It is a matter of belief or ideology how we choose to think about the universe, and it will make no difference how we do our science. ... Science remains a neutral way of explaining things, not anti-God or atheistic. ... I do believe, however, that religious belief can explain more than unbelief can do."[37]

Another prominent scientist who wrote about his religious belief is

---

37. Owen Gingerich: *God's Universe*, pp. 73 ff., passim.

Francis Collins, the leader of the International Human Genome Project. In his book *The Language of God* he writes: "For me the experience of sequencing the human genome, and uncovering this most remarkable of all texts, was both a stunning scientific achievement and an occasion of worship. Many will be puzzled by these sentiments, assuming that a rigorous scientist could not also be a serious believer in a transcendent God. This book aims to dispel that notion, by arguing that belief in God can be an entirely rational choice, and that the principles of faith are, in fact, complementary with the principles of science." [38]

As to be expected, there are the believers and the non-believers, and every shade in between. For example Alister McGrath, a theologian with a scientific training, is an absolute believer: "I write as a Christian, who holds that the face, will and character of God are fully disclosed in Jesus of Nazareth." Among the Roman Catholics, Claude Tresmontant writes: "The Church has been communicating throughout the centuries the creative information which she has received initially from her Lord, to the whole of humanity which it is her task to transform." Jean Delhaye has reservations: "As to me, I belong to the Catholic Roman Church. I am thankful for all the riches I own to her, and the tie which attaches me to her is a tie of love. But I regret her to heavy institutional character, certain archaisms in the formulation of the faith which it is her mission to transmit, a dogmatism of which one does not find in the Gospels, some misuses of a power which she holds without any contestation, and her hesitations in front of any innovation." [39] The mathematician Paul Germain sees his faith as purely personal: "I believe. This has nothing to see with scientific reasoning. It is an affirmation which is my own, which I take as my own responsibility. I see myself as part of the Church and I commit myself to her. I bet my life on her, freely and daily, without any rational proof and without experimental evidence." [40]

The Unknown God

If scientists admit that they know so little about religion and God, why are they talking so much about him, and this in statements which

---

38. Francis Collins: *The Language of God*, p. 3.
39. Jean Delumeau (ed.): *Le savant et la foi*, p. 40-1.
40. id., p. 102.

read like final verdicts? "I simply do not know enough about religions to write with any confidence about them," we remember Daniel Dennett confessing, but he composes all the same an ample polemical volume on the topic. "Imagine someone holding forth on biology whose only knowledge of the subject is the *Book of British Birds*, and you have a rough idea of what it feels like to read Richard Dawkins on theology," finds Terry Eagleton.[41] As we have seen, the 'God' in the mind of practically all scientific materialists is the Judeo-Christian God, the irresponsible autocrat seated on the clouds. As to their knowledge of religion, it is in most cases limited to anecdotes they have heard in their youth: the horrors and bizarreries in the Old Testament, the unbeliev-able miracles in the New Testament, the tragi-comedy of Christianity's history, and some quotations from Voltaire (who was, though not reli-gious, an ardent believer in God), David Hume and Bertrand Russell.

The physicist Paul Davies started his career as a bestselling science author with *God and the New Physics* (1983) and *The Mind of God* (1992). For the second book he was awarded the Templeton Prize for Progress in Religion to the amount of roughly $1 million, slightly above the value of the Nobel Prize. It is the most lucrative prize in the world today, prestigious enough to be presented in Westminster Abbey by royalty. In this book Davies writes: "I would rather not believe in supernatural events personally. Although I obviously can't prove that they never happen, I see no reason to suppose that they do. My inclination is to assume that the laws of nature are obeyed at all times. But even if one rules out supernatural events, it is still not clear that science could in principle explain everything in the physical universe. ... The 'ultimate' questions will always lie beyond the scope of empirical science as it is usually defined."[42] Reading both books, one finds them void of any knowledge about religion, except for the usual clichés. The prize may have been awarded because a scientist deigned to touch the subject of religion at all. In *God and the New Physics* Davies had already written: "In my opinion science offers a surer path to God than religion."[43]

The professional knowledge of the scientists who attack religion is poorly balanced by their ignorance in religious matters. Carl Sagan has given an example of the formation required of them. "Imagine you seri-ously want to understand what quantum mechanics is about. There is a mathematical underpinning that you must first acquire, mastery of each

41. Alister McGrath: *The Dawkins Delusion*, p. 4.
42. Paul Davies: *The Mind of God*, p. 15.
43. Paul Davies: *God and the New Physics*, p. ix.

mathematical subdiscipline leading you to the threshold of the next. In turn you must learn arithmetic, Euclidean geometry, high school algebra, differential and integral calculus, certain special functions of mathematical physics, matrix algebra, and group theory. For most physics students, this might occupy them from, say, third grade to early graduate school – roughly fifteen years. Such a course of study does not actually involve learning any quantum mechanics, but merely establishing the mathematical framework required to approach it deeply."[44]

How is it then that the same scientists, thoroughly schooled in their profession and indoctrinated in their reductionist worldview, think they are qualified in everything else, including the fundamental matters of human existence? "Has there ever been a religion with the prophetic accuracy and reliability of science?" asks Sagan.[45] The predictive accuracy of physics, and of no other science, is limited to physical processes within exactly repeated experimental circumstances. If scientific formulas and mechanisms can be approximately repeated in the complexity of the world as given, it is thanks to the technological skill of the engineers who have, by trial and error, made them applicable. And it is telling to read about "reliability" decades after Kuhn's essay on the relativity of all scientific paradigms.

Indeed, the miracles one reads about in the religious literature are most often folklore or superstition. On what grounds, however, does science suppose that its own miracles are more acceptable? For it demands that you should believe that the whole gigantic universe originated from something much smaller than a mustard seed; that most of your body and the bricks of your house are empty space; that the smallest bit of matter is at the same time a particle and a wave; that matter is nothing but energy although you may badly butt your head against the door; and, yes, that the Sun moves around the Earth, despite you seeing otherwise with your own eyes every day. What is more, each one of the theories by which these wonders are explained is being questioned at the moment. Of the "accuracy and reliability" of the biological sciences, we have had some glimpses earlier.

"No information supposedly gained during a mystical or religious experience, which could not have been otherwise known to the individual claiming the experience, has ever been confirmed," asserts Victor Stenger. Probably not in the literature he has read, which will not have

---

44. Carl Sagan: *The Demon-Haunted World*, p. 237.
45. id., p. 33.

included, let us say, the writings of the medieval mystics of the Rhine Valley or the Zen masters. "All spiritual disciplines, in the East and in the West, have a common core of experience," wrote Sri Aurobindo,[46] who *had* read the literature of both hemispheres, and who had had such experiences himself. "If a person undergoes a religious experience that truly places her in communication with some reality from beyond the material world," continues Stenger, "then we may reasonably expect that person to have gained some deep, new knowledge about the world that can be checked against the empirical facts." Yet, empirical facts are from the material world, not from beyond it. And if Stenger wants to disprove the existence of God (in *God – The Failed Hypothesis*), he should at least have read something or other about the difference between subjective and objective reality.

"Subjective reality cannot be referred to the evidence of the external senses; it has its own standards of seeing and its inner method of verification: so also supraphysical realities by their very nature cannot be referred to the judgment of the physical or sense mind except when they project themselves into the physical, and even then that judgment is often incompetent or subject to caution; they can only be verified by other senses and by a method of scrutiny and affirmation which is applicable to their own reality, their own nature. There are different orders of reality; the objective and physical is only one order. It is convincing to the physical or externalizing mind because it is directly obvious to the senses, while of the subjective and the supraphysical that mind has no means of knowledge except from fragmentary signs and data and inferences which are at every step liable to error. Our subjective movements and inner experiences are a domain of happenings as real as any outward physical happenings ..." (Sri Aurobindo[47])

"Scientists studying mysticism are still in the fact-accumulation stage, and may always be," is the conclusion of John Horgan's inquiry in *Rational Mysticism*. Here the following quotation from Brian Pippard (1920-2008), the late Cavendish professor of physics at Cambridge University, may be apt: "The scientist is right to despise dogmas that imply a God whose grandeur does not match up to the grandeur of the universe he knows. But when we have chased out the mountebanks there remain the saints and others of transparent integrity whose confident belief is not to be dismissed simply because it is inconvenient and uncharted.

---

46. Sri Aurobindo: *Letters on Yoga*, p. 132.
47. Sri Aurobindo: *The Life Divine*, p. 648-9.

We may lack the gift of belief ourselves, just as we may be tone-deaf, but it is becoming in us to envy those whose lives are radiant with a truth which is no less true for being incommunicable. As scientists we have a craftsman's part to play in the City of God; we cannot receive the freedom of that city until we have learnt to respect the freedom of every citizen."[48]

To believe is proper to the human personality; without belief, or a mosaic of beliefs, he cannot live. Here we find again that the essential question and target of the positivists is not religion as such, but the constitution of the human species (to which they too belong), more specifically the human mind. As Jacques Arsac puts it: "One cannot escape believing. Either one believes that there is nothing exterior to science, or one believes that there is something. It is not science which answers this question, it is not science which accepts or refutes that 'exterior', for a [mathematical] formal system cannot say anything about what is exterior to it. This is a question which is of the order of belief, and nobody can escape it. Every answer is a way of reading, of interpreting the book of science, and belongs to philosophy"[49] – which ultimately, in its origin and its ground, belongs to the realm of the spirit.

A Diminished God

"The God whose existence Dawkins is prepared to challenge seems a curiously diminished figure," writes David Berlinski.[50] One expression of Dawkins' idea of God is as follows: "Any designer capable of constructing the dazzling array of living things would have to be intelligent and complicated beyond all imagining. And complicated is just another word for improbable ... Either your God is capable of designing worlds and doing all the other godlike things, in which case he needs an explanation in his own right. Or he is not, in which case he cannot provide an explanation."[51]

What Dawkins and Co. do not seem to realize is that the concept of 'God' they are juggling with is of the same order as their own very human dimensions. They are attacking the shadow of Yahweh, yet the archetype determining their reasoning is identical. It is a 'God' in the

---

48. Kitty Ferguson: *The Fire in the Equations*, p. 260.
49. Jacques Arsac: *La science et le sens de la vie*, p. 220.
50. David Berlinski: *The Devil's Delusion*, p. 150.
51. Richard Dawkins: *Climbing Mount Improbable*, p. 68.

image of man, tribal because Eurocentric, and quite childish. Their arguments hardly differ from those already in use in the popular, exoteric religions of Egypt, Greece, and Rome. If an atom and a cell are so fantastically complex as science has discovered; if there are trillions of cells active in a human body, and trillions of planets, stars, galaxies, quasars and black holes in the universe, all in unceasing motion; if trillions of particles are traversing the Earth and every body on it, every moment of their existence – what being in the image of man, even magnified, could keep this show going in all its tiniest and most gigantic parts?

There is no excuse for the professed ignorance of scientists and philosophers of the subject on which they are not only writing extensively, but intend to effect an adjustment in the opinions (the discourse) of humankind. Such a way of argumentation, at times heightened into raving, is at the very least unscientific, often logically inaccurate if not internally contradictory, and uncultured for its lack of factual and historic knowledge. Moreover, the arguments of most authors in this field are a beguiling potluck of theories and hypotheses which true science keeps apart with care and of necessity. Astronomy is different from quantum field theory, plasma research from paleontology, anthropology from herpetology – without even considering the 'human sciences'. Evolution is a long-term historical event, having taken and taking place one time without any element ever repeating itself. If such, it can only be documented and never proven, for scientific proof demands repetition within similar conditions.

The reason of these reflections is not to be pedantic, but to adduce the necessary thought for the consideration of an evaluation one finds time and again in the writings of the anti-religious: 'God' as a tinkerer or a fumbler. To quote a few such remarks: "An intelligent designer might have managed without these chaotic episodes of boom and bust [like the Cambrian explosion and the Permian extinction]. (Hitchens) ... The design of the eye is not just bad design, it is the design of a complete idiot. (Dawkins[52]) ... There is certainly a lot of order in the universe, but there is also a lot of chaos. The centers of galaxies routinely explode, and if there are inhabited worlds and civilizations, they are destroyed by the millions, with each explosion of the galactic nucleus or a quasar. That does not sound very much like a god who knows what he, or she, is doing. (Sagan) ... If God created the universe as a special place for humanity, he seems to have wasted an awfully

---

52. Richard Dawkins: *The Greatest Show on Earth*, p. 354.

large amount of space where humanity will never make an appearance. ... He wasted a lot of time too. ... Let us also ponder the enormous waste of matter. (Stenger)"

The amazing pretension behind such statements is that the authors seem to know what God is, intends and does – while, being human, they doubtlessly have some problems left with understanding themselves. If they have anything to do with the search for truth which is science, or with the search for truth which is philosophy, why do they not limit themselves to the findings they can validate? Agreed, the discussion of our limited knowledge within a larger framework is an innate exercise of the human mind, as illustrated abundantly throughout history, and discussion is part of the scientific process. But there is a difference, of sincerity or intelligence, between discussion, speculation and condemnation, or just talking one's head off without rime or reason, repeating common platitudes.

"With a necessary part of its collective mind, religion looks forward to the destruction of the world," writes Christopher Hitchens. "The sun is getting ready to explode and devour its dependent planets like some jealous chief or tribal deity."[53] This is an amazing gaffe. Certainly, the Abrahamic religions live in the expectation of the end of the world by water or fire, prelude to the Last Judgment. But the heath death of the Earth preceding the death throes of the Sun is one of the favorite themes of popular science, much rehashed by the media in holiday seasons when there is not much else to report. "If you accept the ordinary laws of science, you have to suppose that human life and life in general on this planet will die out in due course: it is merely a flash in the pan; it is a stage in the decay of the solar system." (Bertrand Russell[54])

The Earth is going to die. When? In five *billion* years. (Mankind is supposed to have originated around two *million* years ago.) "The surface of the Sun at a temperature of several thousand degrees will come extremely close. The Earth will be charred; it will be a cinder." The Sun is going to die. "The sun that has nurtured us for so long will turn into a violent and unpredictable monster prone to sudden nuclear shutdowns and re-ignitions, expansions and contractions, alternating over several millennia. It will end its days as a red giant so gigantic that is wayward mantle of gases will swallow up half the planets of the solar system, including the Earth." The universe is going to die. (The universe is

---

53. Christopher Hitchens: *God is not Great*, pp. 65, 94.
54. *Russell on Religion*, p. 82.

nowadays thought by most cosmologists to have begun 14 billion years ago. Its future life is estimated at about 100 trillion years.) "The stars begin to fade like guttering candles and are snuffed out one by one. Out in the depths of space the great celestial cities, the galaxies, cluttered with the memorabilia of ages, are gradually dying. Tens of billions of years pass in the growing darkness. Occasional flickers of light pierce the fall of cosmic night, and spurts of activity delay the sentence of a universe condemned to become a galactic graveyard" …

It has already been noted in a former chapter that positive science seems to relish demeaning the human condition – when it is not promoting itself by praising "the grandeur of its view of life." The Copernican Principle, or Principle of Mediocrity states that humans are nothing special in the universe – though more and more voices are heard which assert the contrary, and which find support in the anthropic principle. The second law of entropy, the law of inevitable degradation, is the staple reference of scientific materialism – but no organism is an independent or closed system, and now the universe itself is thought by many physicists to be part of a greater whole. It is with gusto that the representatives of 'black science' state and repeat on all possible occasions that humans are animals – although, when it suits their view, they will declare them 'special animals', whose constitution and functioning remains for the greatest part unknown. "By taking the Darwinian 'cold bath', and staring a factual reality in the face, we can finally abandon the cardinal false hope of the ages: that factual nature can specify the meaning of our life by validating our inherent superiority, or by proving that evolution exists to generate us as the summit of life's purpose," writes Carl Zimmer.[55] Stephen Gould denies any sense of progress in evolution, and evaluates bacteria as more successful than humans.

All the same, it would be wrong to include *all* scientists in the ranks of 'black science'. There are some who are as questioning and open-minded as those of the great generation in the first half of the 20th century, and their number may increase because of the present crisis in physics and genetics. In the history of the Gifford Lectures as studied by Larry Witham, " the scientists, philosophers, and theologians brought many views of God to bear, but when it came to find God in nature, there were two primary options: a process God or a traditional omniscient monarch: a God who grew up with the universe or a king

---

55. Carl Zimmer: *Evolution*, p. xvi.

who imposed his will on nature. A third kind of deity, which was more Platonic, was a borderline case, but was generally viewed as transcendent, as far outside of nature, whether as an idea, mathematical form, the ground of being, or 'the wholly other'."[56]

Witham quotes Freeman Dyson as having said: "God learns and grows as the universe unfolds," which is close to Sri Aurobindo's "progressively manifesting god." This idea Sri Aurobindo quotes also in the original Greek of Heraclitus, "*ho theos ouk estin alla gignetai*": God is not but he is becoming, which here means developing, increasing or growing in the evolution on Earth, as born out by the increasing level of consciousness. Dyson said he found this viewpoint "congenial and consistent with scientific common sense. I do not make a clear distinction between mind and God. God is what mind becomes when it has passed beyond the scale of our comprehension."

Few words are more often quoted that those of Stephen Hawking at the end of his *Brief History of Time*, where he expresses the hope that a complete theory of physics, a Grand Unified Theory, will soon be discovered. If that is found, "it would be the ultimate triumph of human reason – for then we would truly know the mind of God." Opinions of the tragic case of mathematical genius that is Stephen Hawking differ, especially about the value of his theories. His quoted words, however well formulated and strategically placed to conclude his book, contribute to the fallacy that the mind of God would be of the order of the human mind, albeit of the most intelligent among *homines sapientes*. Mathematicians have marveled at the complexity, refinement and beauty of mathematics, and seen God as a Mathematician. Others have called him the Great Geometer or the Great Architect. But no reasonable person could suppose that a galaxy and a grain of sand, or a living cell, are held together, in their unimaginable complexity, by a mind comparable to the human. If God is omniscient and omnipotent, the divine Mind must be of a different order, it must be a 'supermind', which is a word, a label covering by definition something of which we can have no idea.

"We have to regard therefore this all-containing, all-originating, all-consummating Supermind as the nature of the Divine Being, not indeed in its absolute self-existence, but in its action as the Lord and Creator of the worlds. This is the truth of that which we call God. Obviously, this is not the too personal and limited Deity, the magnified

---

56. Larry Witham: op. cit., p. 247.

and supernatural Man of the ordinary occidental conception, for that conception erects a too human eidolon of a certain relation between the creative Supermind and the ego ... Supramental nature sees everything from the standpoint of oneness and regards all things, even the greatest multiplicity and diversity, even what are to the mind the strongest contradictions, in the light of oneness; its will, ideas, feelings, sense are made of the stuff of oneness, its actions proceed upon that basis. Mental nature, on the contrary, thinks, sees, wills, feels, senses with division as a starting point and has only a constructed understanding of unity; even when it experiences oneness, it has to act from the oneness on a basis of limitation and difference." (Sri Aurobindo[57])

Omnipresent Reality

"Generally speaking, when we have no evidence or other reason for believing in some entity, then we can be pretty sure that entity does not exist. We have no evidence for Bigfoot, the Abominable Snowman, and the Loch Ness Monster, so we do not believe we exist. If we have no evidence or other reason for believing in God, then we can be pretty sure that God does not exist," writes Victor Stenger.[58] The evidence he has in mind is obviously of the scientific, reductionist kind, valid only where material objects are the case. David Berlinski reasons to the contrary: "Either the Deity is a material object or he is not. If he is, then he is just one of those things, and if he is not, then materialism could not be true."[59] What kind of God would it be who is a material object perceptible by the human senses, and as such the potential object of scientific experimentation?

No evidence? Only if one is blind or refuses in principle to look at the facts, like there were some who refused to look through Galileo's telescope. There is, for instance, Arthur Huxley's *The Perennial Philosophy*; or the literature of great Christian mystics like Meister Eckhart, Margarete Porete, Hadewych, St. John of the Cross; or the heritage of the Zen Masters, whose poetry is truth and truth poetry; or the Indian mystics; or the *Ramayana* and *Mahabharata*, the most voluminous epics in the world, filled to the brim with spiritual lore. John Horgan has the

57. Sri Aurobindo: *The Life Divine*, pp. 132, 965.
58. Victor Stenger: op. cit., p. 18.
59. David Berlinski: *The Devil's Delusion*, p. 53.

following anecdote about Bernard McGinn, theologian and author of a multi-volume history of Christian mysticism: "When I asked McGinn if he had personal acquaintance with hellish or heavenly mystical states, he chuckled uneasily and replied: 'I tend to try to stay away from that.'"[60] But mysticism is a matter of experience, and so is the evidence of God.

The authentic spiritual experience is the same in East and West, and it is the same at the core of every religion, for the human being is the same everywhere. "The perennial philosophy holds that the world's great spiritual traditions, in spite of their obvious differences, express the same fundamental truth about the nature of reality, a truth that can be directly apprehended during a mystical experience." (John Horgan[61]) "The fundamental truth of spiritual experience is one, its consciousness is one," writes Sri Aurobindo, "everywhere it follows the same general lines and tendencies of awakening and growth into spiritual being, for these are the imperatives of the spiritual consciousness. But also there are, based on those imperatives, numberless possibilities of variation of experience and expression: the centralization and harmonization of these possible, but also the intensive sole following out of any line of experience are both of them necessary movements of the emerging spiritual Conscious-Force within us. Moreover, the accommodation of kind and life to the spiritual truth, its expression in them, must vary with the mentality of the seeker so long as he has not risen above all need of such accommodation or such limiting expression."[62]

Still, even in the camp of the anti-religious writers the new spirit of the inner exploration as opposed to the outer, mental and dogmatic one, seems to have penetrated. Sam Harris, in his book *The End of Faith*, writes: "Mysticism is a rational enterprise. Religion is not." He writes about a wealth of mystical evidence. "Spirituality can be – indeed, must be – deeply rational, even as it elucidates the limits of reason. ... When the great philosopher mystics of the East are weighed against the patriarchs of the Western philosophical and theological traditions, the difference is unmistakable: Buddha, Shankara, Padmasambhava, Nagarjuna, Longchenpa, and countless others down to the present have no equivalent in the West. In spiritual terms, we appear to have been standing on the shoulders of dwarfs. It is little wonder, there-fore, that many Western scholars have found the view within rather

---

60. John Horgan: *Rational Mysticism*, p. 41.
61. id., p. 17.
62. Sri Aurobindo: *The Life Divine*, p. 887.

unremarkable."[63] Indeed, as Sri Aurobindo observed almost a century ago: "To this mutual self-discovery and self-illumination by the fusion of the old Eastern and the new Western knowledge the thought of the world is already turning."

The necessity of Something which brings forth the world, life and ourselves, with all the marvels, riddles and atrocities, cannot but keep humanity spellbound, now as it has in times past. Many are the myths which have tried to explain how it all came about and what it all means, for a human's understanding is limited, as are his or her powers, and the fear for tomorrow is ever present. In the myths the human capacities were magnified to the superhuman and mysterious, and that was 'God'. The rationally thinking people in the West can no longer accept this kind of projection, but they have enclosed themselves in their own mind, proud of their rationality and of the new world it has produced, and hanging on to matter as to a raft on a stormy sea.

God, the Divine, 'That' has to be something more, essential, worthwhile, great. The patient people of the East discovered 'That' long ago, but the discovery is difficult and may demand the wager of an entire life. Above the mind one has to reach the spiritual levels of the being, behind the heart one has to enter the soul. There one finds what one really is – and what all is. "Brahman is in all things, all things are in Brahman, all things are Brahman," said those who had attained the goal.

"An omnipresent Reality is the truth of all life and existence whether absolute or relative, whether corporeal or incorporeal, whether animate or inanimate, whether intelligent or unintelligent, and in all its infinitely varying and even constantly opposed self-expressions, from the contradictions nearest to our ordinary experience to those remotest antinomies which lose themselves on the verges of the Ineffable, the Reality is one and not a sum or concourse. From that all variations begin, in that all variations consist, to that all variations return. All affirmations are denied only to lead to a wider affirmation of the same Reality. All antinomies confront each other in order to recognize one Truth in their opposed aspects and embrace by the way of conflict their mutual Unity. Brahman is the Alpha and the Omega. Brahman is the One besides whom there is nothing else existent." (Sri Aurobindo[64])

---

63. Sam Harris: *The End of Faith*, p. 215.
64. Sri Aurobindo: op. cit., p. 33.

14.

Intelligent Design

What stark Necessity or ordered Chance
Became alive to know the cosmic whole?
What magic of numbers, what mechanic chance
Developed consciousness, assumed a soul?

SRI AUROBINDO

The simplicity that was once expected to be the foundation of life has
proven to be a phantom; instead, systems of horrendous, irreducible
complexity inhabit the cell. The resulting realization that life was
designed by an intelligence is a shock to us in the twentieth century.
But other centuries have had their shocks, and there is no reason to
suppose that we should escape them.

MICHAEL BEHE

In the last decades of the 20th century two new ways of thinking have put scientific materialism into question from within. Its premise, to wit that all is matter because there is nothing but matter, is a dogma and the ground of a hardened paradigm. It is therefore by no means surprising that the present twofold viral infection in the innards of science itself has been countered by strong defence mechanisms. The controversies are still ongoing and may continue for a long time to come. One new idea is called 'the anthropic principle', the other 'intelligent design theory'.

The Anthropic Principle

The anthropic principle has originated from the physicists' wonderment about the improbability of the existence of conscious life – in other words: of ourselves – on Earth. Contemplating the glories of nature, and the human in nature, has caused delight everywhere and in all climes, and is at the heart of natural philosophy. The anthropic principle, however, is the result of modern, in fact quite recent macro- and microphysical findings. "The anthropic principle is essentially the idea that our very existence puts constraints on what physical laws are possible. These must be such that intelligent beings such as ourselves could somehow evolve." (Peter Woit[1]) It has become incontestable that "life can arise only in an extremely narrow range of all possible physical parameters and yet, oddly enough, here we are, as though the universe had been designed to accommodate us." (Lee Smolin[2])

In the history of the universe there is a range of mathematical constants which are absolutely basic for its evolution. There are for instance the fundamental physical constants which no theory can explain, but without which no theory can work: Planck's constant, the speed of light, the electrical charge of the proton, Newton's constant of gravity, the mass of the electron, and several others. The constants have been described as "the portions, balances, and strengths by which particles and forces worked together to create matter, elements, chemistry, and finally biological life. Thanks to the constants, it seems, the universe was not a cosmic soup of particles, as it might have been. ... Where do the constants come from? They were simply there from the very beginning. They produced what would be called a 'fine-tuned' universe."[3]

Without the fine-tuning of the universe in its eventful history since the Big Bang, it would not be as it is now, or it would have ceased to be in one of the critical phases of its evolution. For instance, the electromagnetic force is 39 orders of magnitude stronger than the gravitational force. Why 39, neither more nor less? There is no answer. Yet "if the forces were more comparable in strength, stars would have collapsed long before life had a chance to evolve. ... The neutron is heavier than the proton, but not so much heavier that neutrons cannot be bound in nuclei, where conservation of energy prevents the neutrons from decaying. Without neutrons we would not have the heavier elements needed for building complex systems such as life."[4] A physicist is of the opinion that the laws of nature seem specially tailored to our own existence; a biologist reflects that if the history of life had been even

---

1. Peter Woit: *Not Even Wrong*, p. 12.
2. Lee Smolin: *The Trouble with Physics*, p. 162.
3. Larry Witham: *The Measure of God*, p. 144.
4. Victor Stenger: *God – The Failed Hypothesis*, p. 146.

slightly different, the makeup of the world's species today would be completely different.

There are actually two kinds of anthropic principle, differing in the value they confer on the human being and its place in the universe. The following formulations are by Brandon Carter, the astrophysicist who coined the term in 1973. The weak anthropic principle: "We must be prepared to take account of the fact that our location in the universe is necessarily privileged to the extent of being compatible with our existence as observers." 'Location' here means our position on the Earth in space and time. The strong anthropic principle: "The universe, and hence the fundamental parameters on which it depends, must be such as to admit the creation of observers within it at some stage."[5] The observer is a living, conscious being. According to the weak principle this being may passively state the fact that the universe has created the (very improbable) circumstances of his being there; the strong principle asserts that all elements in the evolution of the universe *must* lead to the creation of the conscious observer.

It will cause no surprise that statements of this kind have led to vehement protests from scientific materialists, according to whom matter is not capable of intention and all events in the universe are chance events. Still, the improbability of the evolution of the universe leading up to the appearance of a being like the human is of such order of magnitude that even confirmed materialistic scientists, and among them some of the greatest, agree that something more than chance must have been in play. "The more I examine the universe and study the details of its architecture, the more evidence I find that the universe in some sense must have known we were coming," concedes Freeman Dyson. And even Steven Weinberg enunciates his amazement that the laws of nature and the initial conditions of the universe should allow for the existence of beings who could observe it. "Life as we know it would be impossible if any one of several physical quantities had slightly different values."

"Until very recently, the anthropic principle was considered by almost all physicists to be unscientific, religious, and generally a goofy misguided idea. According to physicists it was a creation of inebriated cosmologists, drunk on their own mystical ideas," writes Leonard Susskind.[6] "Many scientists dislike the anthropic principle because it seems to be a throwback to a pre-Copernican, Aristotelian style of reasoning.

---

5. http://en.wikipedia.org/wiki/anthropic_principle
6. Leonard Susskind: *The Cosmic Landscape*, p. 14.

It seems to imply an anthropocentric view of the cosmos." (Freeman Dyson[7]) "It explicitly invokes life and is consequently anathema to many scientists. ... The anthropic principle re-injects teleology into science." (George Sudarshan[8]) These are the reasons why some sarcastic scientists launched, besides the weak and the strong anthropic principle, a third variety, "the completely ridiculous anthropic principle"!

Nonetheless, the data now at the disposal of the scientists of the micro- and the macroworlds, as those of biology, differ enormously from the data that were available to a Galileo, Descartes or Newton. The principle of studying the facts as perceived by the senses may remain the same, but – and this is only one different factor – instruments have extended the perception of the senses in a way scientists of the previous centuries could not so much as imagine. The use of the microscope and the telescope for scientific purposes caused a revolution in Galileo's and Leeuwenhoek's time. The modern instruments in the micro- and the macroworld are causing another revolution which is not less important.

Secondly, the fundamental significance of the levels of existence – matter, life, mind – will have to be recognized in the near or further-off future, because such is reality. New interpretations or a new paradigm will be needed to approach life, which is not reducible to measurement, calculation and mass only. 'Vitalism' has refused to go away in the history of science; at one time it will have to be accepted and studied, because it will become apparent that matter alone does not suffice to explain reality. And its study will be indispensable for the knowledge and mastery of consciousness in a following stage of humanity's progress. Such will be the science of tomorrow which is already knocking at today's door in issues like the anthropic principle and 'intelligent design', to which we now turn.

Intelligent Design and the Other Evolutionary Theories

The 'Intelligent Design' phenomenon is generally called 'the Intelligent Design theory', which is a misnomer. Intelligent Design puts scientific materialism into question without as yet being able to construct its own, rival theory. For it acknowledges life and intelligence, or consciousness, and, as we have just seen, a systematic explanation of

---

7. Freeman Dyson: *Infinite in all Directions*, p. 296.
8. Tony Rothman and George Sudarshan: *Doubt and Certainty*, p. 234.

this elementary categories is not yet possible. This is an observation of note because Intelligent Design (ID) is often attacked for its shortcomings as a theory, which it is not, and because adherents of ID themselves sometimes try to present it as such. The question formulated by ID is well-founded on scientific grounds; the answer remains to be found, and will require an entirely new approach of science.

Scientific materialism, inherently limited like all constructions of the mind, has been criticized since its beginnings in the 17th century. It gradually became the commonly accepted and academic interpretation of nature and reality in the 19th century. Curiously, while its foundations were again seriously questioned in the physics of the first half of the 20th century, it triumphed in the biological sciences with the discovery of the double helix in 1953 and the centenary of the publication of Darwin's *Origin of Species* in 1959. Although biology relied on physics as its nethermost basis, the background mentality in the two fields was quite dissimilar, physics having its tradition of the great "mystics" (Galileo, Kepler, Newton, and the Einstein-Bohr generation), biology convinced of its metaphysical truths within a chitin shell of dogmatism, possibly secreted because of sheer vulnerability.

Because of the Second World War science's centre of gravity shifted from Europe to the United States. The constitution of this country and its Bill of Rights are based on the principles of the Enlightenment transmitted via the humanitarian ideals of Freemasonry. On the other hand, there is still the widespread influence of the Christian sectarian refugees who were its first settlers, the emotional religiosity of Bible thumping preachers, and Christian religious fundamentalism in general. These two stances, rationalism and religious fundamentalism, are the great divide that runs through the nation and heavily determines its politics, culture and religious activities.

One finds this internal division also expressed in the trials about the teaching of either evolution or creationism, or both, in educational institutions. The most famous remains the "Monkey Trial" of 1925, but it has been followed by several others, like *McLean v. Arkansas* in 1982, and *Edwards v. Aguillard* in 1987. It was on this occasion that a group of scientists, including Nobel laureates, published in the newspapers a manifesto in support of scientific materialism. Other "court defeats for creationism" occurred in 1990, 1994 and 1997. At first a typical American phenomenon, such controversies have now spread to other countries, sometimes provoked not by Christian but by Islamic creationists. The contentious atmosphere resulting from these debates is directly responsible for the bitterness of the battles fought between creationists, positivists and intelligent-designers.

Michael Denton's *Evolution: a Theory in Crisis* (1985) may be seen as the starting signal of the Intelligent Design movement. On its cover is printed: "New developments in science are challenging orthodox Darwinism." Denton had no religious program; his criticism of Darwinism, which in most cases meant neo-Darwinism, rested on purely scientific grounds and is generally respected, although it came as a shock after the triumphalist Darwinian wave in the 1950s and 1960s, Desmond Morris' *Naked Ape* and Jacques Monod's *Chance and Necessity*. "I have tried to show why I believe that the problems [of 'Darwinism'] are too serious and too intractable to offer any hope of resolution in terms of the orthodox Darwinian framework, and that consequently the conservative view is no longer tenable," wrote Denton.[9] His was a strong challenge indeed.

"Darwin's model of evolution is still very much a theory and still very much in doubt when it comes to macro-evolutionary phenomena," Denton asserted. "Furthermore, being basically a theory of historical reconstruction, it is impossible to verify by experiment or direct observation as is normal science. ... Unique events are unrepeatable and cannot be subjected to any sort of experimental investigation. ... Not only is the theory incapable of proof by normal scientific means, the evidence is far from compelling."[10] Denton, an Australian molecular biologist, then went on, in an unusually cultured book, to substantiate his contentions by examining the main problems such as the dogmatism of 'Darwinism', the proof of homology, the fossil record, the importance of the molecular biological revolution, and the enigma of life's origin.

These solid arguments from a neutral source were grist to the mill of several Christian intellectuals who accepted some version of evolution but not the materialistic dogma buttressing Darwinism. Initially the most vocal one was Phillip Johnson, professor of law and Christian convert, who published *Darwin on Trial* in 1991. The reader may remember that, by then, the sociobiologists Edward Wilson and Richard Dawkins had published their supposedly 'arch-Darwinian' and aggressively anti-religious, not to say anti-theistic, books. Johnson saw sociobiology in particular and scientific materialism in general as a lethal menace to

---

9. Michael Denton: *Evolution: a Theory in Crisis*, p. 16.
10. id., p. 75-6.

Western culture, by him identified with the Christian religion and morals. He was soon followed in this by like-minded but scientifically qualified persons who created organizations and published a literature to combat the institutionalized threat of official science.

"On the one hand, modernists say that science is impartial fact-finding, the objective and unprejudiced weighing of evidence," writes Johnson. "Science in that sense relies on careful observations, calculations, and above all repeatable experiments. That kind of objective science is what makes technology possible, and where it can be employed it is indeed the most reliable way of determining the facts. On the other hand, modernists also identify science with naturalistic philosophy [Johnson's term for scientific materialism]. In that case science is committed to finding and endorsing naturalistic explanations for every phenomenon – *regardless of the facts*. That kind of science is not free of prejudice. On the contrary, it is *defined* by prejudice. The prejudice is that all phenomena can ultimately be explained in terms of purely natural [i.e. material] causes, which is to say unintelligent causes."[11]

It was Johnson's outspoken intention to go on the counteroffensive against scientific materialism, or as he put it "to split the foundations of naturalism" by "the wedge of truth" (title of one of his books) – Christian truth, that is. He became referred to as 'Phillip "the Wedge" Johnson'. He and his like-minded crusaders saw themselves as defenders of Western civilization, "the normative legacy of Judeo-Christian ethics". From a pamphlet, known as "the Wedge document" we quote the following: "The proposition that human beings are created in the image of God is one of the bedrock principles on which Western civilization was built. ... Thinkers such as Charles Darwin, Karl Marx and Sigmund Freud portrayed humans not as moral and spiritual beings, but as animals or machines. ... The materialist conception of reality infected virtually every area of our culture ... The cultural consequences of this triumph of materialism were devastating ... Discovery Institute's Centre for the Renewal of Science and Culture [a powerhouse of Intelligent Design] seeks nothing less than the overthrow of materialism and its cultural legacies."[12]

The battle was on – and is still on, in dust clouds of confusion. Original Darwinism and subsequent neo-Darwinism; creationism and Christian evolutionism; science as theory and science as metaphysical

---

11. Phillip Johnson: *The Wedge of Truth*, p. 14 (italics in the text).
12. John Forster, Brett Clark and Richard York: *Critique of Intelligent Design*, p. 35.

dogma; Christian theology and Intelligent Design; creationist ID and neutral ID – it is all mixed, together with the historical variations of each of these religions, doctrines, theories, hypotheses, ideas, guesses, or fantasies. The ignorance of the way in which the media present and mispresent this mixture is sometimes atrocious. To create some clarity in the confusion should be worth our while.

Creationism

Creationism is the belief that the world originated through an act of God as described in the Bible, a collection of books holy to the early Hebrew tribes. The Bible was, and still is, a sacred text of three major religions – in historical order: Judaism, Christianity and Islam – although in each religion it has been complemented with additional revelations, doctrines and traditions. The first words of its first book, *Genesis*, are: "In the beginning God created the heavens and the earth," and it goes on describing how God created the universe and everything in it in six days, after which he rested on the seventh. A story in the biblical history of humanity, often referred to in the evolution controversies, narrates how God in his anger with a corrupt humanity causes a flood to destroy everything on Earth, except for the righteous Noah, his family, and a sample of all animal species in his ark. This flood, *the* Flood, is often used to explain the findings of geology and paleontology.

Religious scientists who accept creation as truth "no more think alike than do evolutionists." There are the literalists, for whom everything happened exactly as written in the Bible. Their belief is called "the young Earth interpretation" because it holds that the creation of the Earth, and all things on it, happened not more than a few thousand years ago. The Anglican archbishop James Usher calculated that the year of creation had been -4004. Printed in the margins of the King James Bible, Usher's chronology became quasi gospel for British and American Protestants during the eighteenth and nineteenth centuries. It is still the belief of millions of people who have been taught the biblical story of creation in their early youth.

Other religious scientists, unable to deny the geological and paleontological data but wanting to remain within their faith, have widened the *Genesis* story into a "day-age theory." This means that they take the six "days" of creation as symbolical for six 'ages' of undetermined duration. Their main argument is that, indeed, the biblical story reflects in broad outlines the emergence of the universe, the creation of the

Earth, and the appearance of plants, animals, and humans in the right evolutionary order. The "day-age theory" may be seen as a compromise between religious injunction and the scientific theories of evolution.

A third interpretation of *Genesis* admits "a vast gap of time between creation of the earth and creation of the Garden of Eden," thus again finding accommodation for the modern discoveries of science. And a fourth religious view holds that "God uses evolution as his instrument; evolution is how God creates, his work cannot be detected." This view is already quite distant from a literal acceptance of the biblical creation myth and close to natural theology which sees the hand of God in everything, in the heavens and on Earth. In fact, several Catholic scientists accept *Genesis* as a matter of respect for the doctrines of their Church, *pro forma*, while in the daily practice of their faith they see God as the sustainer of the evolution.

Yet the Christian Bible consists of two collections of books, the Hebrew Old Testament and the Christian New Testament. According to the latter, Christ, the incarnated Son of God, is the key to the redemption of all creation and especially of humankind. Pierre Teilhard de Chardin and his consummation of the evolution in Christ is a well-known example of this vision. Another is e.g. Claude Tresmontant, former professor of philosophy at the Sorbonne, who writes: "Only the revelation and incarnation [of Christ] allow us to discover what is the ultimate aim of the creation." The creation was not complete in the beginning, as one might infer from the Bible, for it cannot be completed without Christ. "In Christ is the meaning of the creation and he realizes in himself the final goal of it. ... The aim of creation is the divinization of the human being. ... The ultimate goal of creation was not the snail, nor the gorilla, nor the australopithecine, it is the True Human united with the True God, the new Human who is born again and has become like Christ." [13]

Positivist Theories of Evolution

'Positivism' and 'reductionism', and in America 'physical naturalism', are terms nearly identical to 'scientific materialism' as used throughout our narrative. Present-day Darwinists routinely present Darwinism as the one and only scientific evolutionary theory, outlined by Charles Darwin in his *Origin of Species*. So does for instance David Attenborough

---

13. Claude Tresmontant: *L'Histoire de l'univers et le sens de la création*, passim.

in his BBC documentary on Charles Darwin, and so do most other popularizations of evolution. This is not only historically incorrect, it is also misleading as to the contents and significance of Darwinism, as it is of evolutionary theory in general. Of this, the reader has been informed in the chapters on Jean-Baptiste de Lamarck and Alfred Wallace.

All three of them, Lamarck, Wallace and Darwin, were exponents of their time, passionate students of their subject, and well-read in all texts available on it, among them the works of Linnaeus, Cuvier, Buffon, Erasmus Darwin and Chambers. Seen in this way, Darwin was much more a synthesizer and a symbol, leaning heavily on the theses of the geologist Lyell and the economist Mathus, than the inventor of scientific evolution. He was ignorant of the natural mechanism which he proposed as a substitute of supernatural creation, simply because in his time such knowledge was not yet available. The idea of evolution was in the air; the new sciences of geology, anatomy, anthropology and paleontology continuously discovered new data supporting it. But what is the scientific value of a theory of which the central mechanisms (variation and natural selection) can only be guessed at, not proved? In fact, the enormous mass of innovations and rearrangements by 'Darwinism' in the last century notwithstanding, the guesswork continues and definitive proof remains to be provided.

At present there are several positivist evolutionary theories, of which 'Darwinism' is the most widespread and by many supposed to be the only one. Lamarckism has been heavily attacked and ridiculed in the past, but it has never completely disappeared and is, as we saw when discussing the ongoing revolution in genetics, gaining strength again. 'Darwinism' is a cluster of theories, of which the original Darwinism is probably the least referred to. What is nowadays understood as Darwinism is generally some kind of neo-Darwinism, and depends on the inclinations of the proponent, whether geneticist, field biologist, anthropopaleontologist, or whatever. The sociobiology of Edward Wilson has radicalized neo-Darwinism into a moral, not to say fundamentalist theory. Dawkins and Gould both proclaim themselves arch-Darwinists, but Dawkins has launched his un-Darwinian theories of the extended phenotype and the meme, and Gould his anti-Darwinian theory of punctuated equilibrium, denying, in blatant heresy, Darwin's principle of gradualism. As André Pichot remarks dryly: nowadays every biologist seems to have his personal theory of evolution.

Another problem is that evolution is a historic science and therefore very different from physics with its prime condition that experimental proof must be repeatable within identical conditions. We remember Michael Denton's observation that "being basically a theory of historical reconstruction, [Darwin's model of evolution] is impossible to verify by experiment or direct observation as is normal science. ... Unique events are unrepeatable and cannot be subjected to any sort of experimental investigation."

Given the limitations of the human mind, even of the greatest, controversy is part of the progress in science, or of the formation and clash of its paradigms. But it may be seen as a sign of its immaturity that it presents each of its new thought systems as absolute truth. The worlds opened to humanity by science are wonderful, its lifting of some corners of the veil of nature beyond expectation, and its guidance in the construction of our global world amazing though disturbing. All this, however, may not be more than a step towards the Truth and Reality of tomorrow.

Intelligent Design Theory

"The simplicity that was once expected to be the foundation of life has proven to be a phantom; instead, systems of horrendous, irreducible complexity inhabit the cell. The resulting realization that life was designed by an intelligence is a shock to us in the twentieth century," writes Michael Behe.[14] We know of the distortions caused by 'mental simplification', at the beginning of modern science in the representation of things, afterwards more and more misrepresenting reality as both the micro- and the macroworld became "horrendously" complicated. Actually, all domains of nature have become so complex, and are at the same time so astonishingly functional, that the question how such complexity may have come about can no longer be disregarded. What or who designed such astonishing, and astonishingly functional, complexity?

"Design is simply the purposeful arrangement of parts,"[15] be they microscopic, small, big, or astronomical. 'Design' here is not a term from the world of fashion or sports shoes. It refers to drafting, planning, purpose, systematic ordering – not by chance, but by an intention which requires an intelligence. "'Design' means the operation of an intelligence that organizes the effects of law and chance, creating

---

14. Michael Behe: *Darwin's Black Box*, p. 253.
15. id., p. 193.

247

information. ... The Intelligent Design theory started to take shape in the late 1970s, as an outcome of information theory. Faced with the enormous complexity of living things, ID theorists argue that it makes more sense to assume that the information is a language [i.e. a coherent system of significance]. In other words, it is the product of design. This is the complete opposite of Darwinism." [16]

The argument for design, as composed by Michael Behe, goes as follows: "1. We infer design whenever parts appear arranged to accomplish a function. 2. The strength of the inference is quantitative and depends on the evidence; the more the parts, and the more intricate and sophisticated the function, the stronger is our conclusion of design." Here, Behe rightly mentions William Paley's old argument that if an artificial object, e.g. a pocket watch (nowadays we would say a chronometer), is found somewhere in nature, the conclusion must be that the watch has been designed by an intelligent being. "3. Aspects of life overpower us with the appearance of design. 4. Since we have no other convincing explanation for that strong appearance of design, Darwinian pretensions notwithstanding, we are rationally justified in concluding that parts of life were indeed purposely designed by an intelligent agent." [17]

Creationist Intelligent Design

Most experts in the ID movement are qualified as experimental or theoretical scientists, or as university professors. Many of them are also religious. Behe, for instance, is a Catholic, and Johnson and Dembski belong to American Christian denominations. Dembski is quoted as having written: "My thesis is that all disciplines find their completion in Christ and cannot be properly understood apart from Christ. ... Christology tells us that the conceptual soundness of a scientific theory cannot be maintained apart from Christ. Christ is the light and the life of the world and its destiny. It follows that a scientist in trying to understand some aspect of the world, is in the first instance concerned with that aspect as it relates to Christ – and this is true regardless of whether the scientist acknowledges Christ." [18]

The militant ID movement has, among other activities, sponsored conferences, one in 1995 titled "Death of Materialism and the Renewal of Culture," another the following year under the title "Mere Creation,"

---

16. Denyse O'Leary: *By Design or by Chance?* pp. 7, 10.
17. Michael Behe: op. cit., p. 205.
18. Niall Shanks: *God, the Devil, and Darwin*, pp. 158-9.

and a third one in 1999, "Science and Evidence for Design in the Universe." In the published lectures of the latter, Bruce Chapman specifies the anti-materialistic objectives of the new movement: "Materialism is not limited in its implications to natural science. Materialism is a way of understanding day-to-day existence and responding to it. ... It can be argued that materialism is a major source of the demoralization of the twentieth century. Materialism's explicit denial not just of design but also of the possibility of scientific evidence for design has done untold damage to the normative legacy of Judeo-Christian ethics. A world without design is a world without meaning."[19] And Phillip "the Wedge" Johnson writes: "If reason is to be a reliable guide, it must be grounded in a foundation that is more fundamental than logic. Instrumental reason is not enough. That is why the fear of the Lord is not the beginning of superstition but the beginning of wisdom."[20]

No wonder that the not-Christian-minded accuse the ID movement of being a cover organization for Christian creationism. As an organization it certainly sees its reason to exist in the fight against gross materialism and for the traditional values of the Christian world. In fact, it is rather amazing that it took so long for such an accumulation of religiously inspired forces *within* science to rise up against the dominant materialism. The violently anti-religious language of authors like Richard Dawkins, which has more to do with his psychological personality than it has with science, certainly contributed to their organization.

Neutral or Open Intelligent Design

The religious ID'ers are of course aware of the arguments that can be directed against them. The biochemist Michael Behe is "a legitimate scientist with a good record of publication," whose book *Darwin's Black Box* (1996) marked a turning point for the new design movement. He writes: "Although some of my critics, noting that I am a Catholic, argue that design is a religious idea, I disagree. I think a conclusion of design is completely empirical, and can be justified solely by physical data, as well as by understanding of how we come to a conclusion of design."[21] He insists that Intelligent Design is a matter of science, and that secondary deductions belong to another domain, whether cultural, religious, or both.

---

19. Foster, Clark and York: op. cit., p.39.
20. Niall Shanks: op. cit., p. 246.
21. William Dembski (ed.): *Uncommon Dissent*, p. 142.

"I am keenly aware," writes Behe, "that in the past few years many people have come to regard the phrase 'intelligent design' as fighting words, because to them the word 'design' is synonymous with 'creationism', and thus opens the door to treating the Bible as some sort of scientific textbook, which would be silly. That is an unfortunate misinterpretation. The idea of intelligent design, although congenial to some religious views of the universe, is independent of them. For example, the possibility of intelligent design is quite compatible with common descent, which some religious people disdain. What's more, although some religious thinkers envision active, continuing intervention in nature [by a higher being], intelligent design is quite compatible with the view that the universe operates by unbroken natural law, with the design of life perhaps packed into its initial set-up."[22]

The cause of the misunderstanding lies of course with a faction of the religious ID'ers themselves, those responsible for the opinions quoted in the previous section. All the same, it must be stressed that Intelligent Design as such, as a scientific statement, or as a questioning perception based on scientific grounds, is a-religious. ID advocates, unlike strict creationists, accept evolution, but they do not accept that it works in the 'Darwinian' way. And "their purpose is not to support *Genesis*, but to follow evidence wherever it leads." In the words of a Phillip Johnson in a neutral mood: "The question for now is not whether the vast claims of Darwinian evolution conflict with *Genesis*, but whether they conflict with the evidence of biology."[23]

In the preface to the papers read at the 1999 "Science and Evidence" conference we find written: "Unlike neo-Darwinists and other evolutionary theorists, design theorists hold that intelligent causes rather than undirected natural [i.e. material] causes best explain many features of life and the universe. Unlike many creationists, design theorists do not necessarily believe that the earth is young, neither do they base their theories upon scriptural texts. Unlike many theistic evolutionists who think design can only be seen through 'the eyes of faith', design theorists believe that scientific evidence actually points to intelligent design – that intelligent design is, in their words, 'empirically detectable'."[24] As William Dembski, mathematician and philosopher, is quoted to have put it: "Intelligent Design is not creationism and it is not naturalism [i.e. materialism]. Nor is it a compromise or synthesis

22. Michael Behe: *The Edge of Evolution*, p. 166.
23. Denyse O'Leary: op. cit., p. 169.
24. Behe, Dembski and Meyer: *Science and Evidence for Design in the Universe*, p. 12.

of these positions. It simply follows the empirical evidence of design wherever it leads. *Intelligent Design is a third way.*"[25]

"The conclusion of intelligent design flows naturally from the data itself – not from sacred books or sectarian beliefs. Inferring that biochemical systems were designed by an intelligent agent is a humdrum process that requires no new principles of logic or science. It comes simply from the hard work that biochemistry has done over the past forty years, combined with consideration of the way in which we reach conclusions of design every day." (Behe[26])

But then: who or what is the agent of that intelligent design? Is it a 'designer' with or without a capital letter? Behe's answer is brief: "Inferences to design do not require that we have a candidate for the role of designer."[27] He comments elsewhere: "The core claim of intelligent design is quite limited. It says nothing directly about how biological design was produced, who the designer was, whether there has been common descent, or other such questions. Those can be addressed separately. It says only that design can be empirically detected in observable features of physical systems."[28]

Let us have a look at these observable features.

Irreducible Complexity

Even Richard Dawkins is baffled by the "horrendous complexity" of life's creatures, for he states laconically in *The Blind Watchmaker*: "The biologist's problem is the problem of complexity."[29] Mental simplification and familiarity with some of its outward aspects may have suggested an understanding of life where as yet there was no more than a rudimentary approximation. In the words of one scientist: "Familiarity dulls our sense of wonder at the craftsmanship of nature." Still, mental simplification and familiarity have both been, and for a long time, the result not only of ignorance but still more of the limitation of our physical senses through which nature is perceived.

As noted before, the means at its disposal have allowed science to penetrate in physical realms previously unimaginable. The complexity

---

25. Denyse O'Leary: op. cit., p. 167 (emphasis added).
26. Michael Behe: *Darwin's Black Box*, p. 193.
27. id., p. 196.
28. Niall Shanks: *God, the Devil, and Darwin*, p. 155.
29. Richard Dawkins: *The Blind Watchmaker*, p. 19.

of the atomic and sub-atomic worlds, of the cells and the microscopic processes of life, and of the cosmic phenomena has been found to be staggering, and led "from complexity to perplexity." "It was once expected that the basis of life would be exceedingly simple," writes Behe. "That expectation has been smashed. Vision, motion and other biological functions have proven to be no less sophisticated than television cameras and automobiles. Science has made enormous progress in understanding how the chemistry of life works, but the elegance and complexity of biological systems at the molecular level have paralyzed science's attempt to explain their origins."[30]

"Each of the anatomical steps and structures that Darwin thought were so simple actually involves staggeringly complicated biochemical processes that cannot be papered over with rhetoric."[31] Marcel Schutzenberger, French mathematician and doctor of medicine, observed: "Whether gradualists or saltationists, Darwinians have too simple a conception of biology. ... In biological reality, the space of even the simplest function has a complexity that defies understanding, and indeed defies any and all calculations."[32]

Behe's comparison of biological functions with television cameras and automobiles is actually far too plain. A hundred years ago the cell was still seen as a minuscule blob of plasma; now "the most elementary type of cell constitutes a 'mechanism' unimaginably more complex than any machine yet thought up, let alone constructed by man."[33] "A cell resembles a miniature industrial complex that is much more complicated than a General Motors or a Boeing plant," writes Geoffrey Simmons in *What Darwin Didn't Know*.[34] A human body, from its birth to its adulthood, contains from 10 to 75 *trillion* cells, all functioning, reacting to each other, and keeping the body alive. The brain is considered the most complex object in the universe. Our small bowel contains an astronomical number of 500 different kinds of bacteria. "How can the sight of a tennis ball's shape, size, colour and speed be sent to dozens of spots in the brain at the same time, be recombined into a functional image, and then result in an action – all in less than a second?"

In stark contrast with the perplexity caused by complexity is the

---

30. Michael Behe: op. cit., p. x.
31. id., p. 22.
32. William Dembski (ed.): *Uncommon Dissent*, pp. 42, 48.
33. Francis Hitching: *The Neck of the Giraffe*, p. 68.
34. Geoffrey Simmons: *What Darwin Didn't Know*, p. 75

self-assuredness of some 'Darwinist' authors in the face of the processes of nature, Richard Dawkins being the prime example. A few quotes must do, but pages could be filled with explanations far from science and close to nonsense. About the origin of life, the great enigma of biology: "Actually a molecule that makes copies of itself is not as difficult *to imagine* as it seems at first, and it only had to happen once. Think of the replicator ... Imagine it as ... Now suppose ..."[35] About natural selection: "If a group of atoms in the presence of energy falls into a stable pattern it will tend to stay that way. The earliest form of natural selection was simply a selection of stable forms and a rejection of unstable ones. There is no mystery about this. It had to happen by definition."[36] About gradualism: "Gradual evolution by small steps, each step being lucky but not too lucky, is the solution to the riddle."[37] What is the science behind "lucky but not too lucky"? Luck is metaphysical speculation; scientific explanations need causes.

"In order to add new segments [to the body of the snake], all that has to be done is a simple process of duplication," teaches Dawkins. "Since there already exists machinery for making one snake segment ... new identical segments may easily be added by a single mutational step."[38] To add any part to the body of any organism is never either easy or simple. For instance: "It is not sufficient to give a bird wings if it is to fly. In addition its bones must be made lighter while at the same time maintaining their strength, feathers must be aerodynamically adapted, the center of gravity must shift, the breastbone and musculature must develop, and changes in metabolism are required to provide sufficient energy for flight. If such changes do not all occur together and in a coordinated fashion, then they may well be disadvantageous to survival."[39]

On this sort of Dawkinsian aberration Stephen Gould comments: "Most so-called explanations amount to little more than what Lewontin and I, following Kipling, would later call "just-so stories" or plausible claims without tested evidence ..."[40] Even Daniel Dennett seems to keep his distance from such kind of happy-go-lucky reasoning and writing: "One may be reasonably nervous about the size of the role of sheer, unfettered imagination in adaptationist thinking." He is aware that "what particularly infuriates Gould and Lewontin is the blithe

35. Richard Dawkins: *The Selfish Gene*, p. 15 (italics added).
36. id., p. 13.
37. Richard Dawkins: *River out of Eden*, p. 97.
38. Richard Dawkins: *The Blind Watchmaker*, p. 290.
39. David Bohm and David Peat: *Science, Order, and Creativity*, p. 202.
40. Stephen Gould: *Punctuated Equilibrium*, p. 3.

confidence with which adaptationists [like Dawkins] go about their reverse engineering," which means *imagining* how organic structures have come about, "always sure that sooner or later they will find *the* reason why things are as they are, even if it so far eludes them."[41]

Considering natures complex processes with an informed but open mind, the conclusion imposes itself that they cannot possibly have come about through Darwinian "gradual evolution by small steps." The mechanisms, processes, organs and organisms cannot have come about by incremental chance processes – for it should not be forgotten that, according to scientific materialism, nature has no awareness, and therefore no plans, no intentions, no teleological vectors. What Darwin imagined as having come about gradually, and is still an article of faith of 'Darwinism', can no longer be accepted in the face of the astonishing complexity revealed by the latest scientific instruments.

The phenomenon of a complexity which can no longer be reduced to its parts and explained by them, has been called 'irreducible complexity'. "By irreducibly complex I mean a single system composed of several well-matched, interacting parts that contribute to the basic function, wherein the removal of any one of the parts causes the system to effectively cease functioning." This is Michael Behe's oft-quoted definition in his book *Darwin's Black Box*.[42] The "systems" are the structures built by life, from the nano-mechanisms in the cell to the gigantic creatures that at one time roamed the Earth in their thousands, to our hyper-complex brain. The systems have one or more basic functions to which all the parts harmoniously contribute, and which cease functioning if one of their parts breaks down or is removed.

The classic example of an irreducibly complex system has become, since the publication of Behe's book, the mousetrap. The basis of a mousetrap consists of a small wooden plank, the 'platform'. On this platform are fixed a spring with a 'hammer', actually a bent metal wire strong enough to kill a mouse; a 'catch', which has to be moved by the mouse to activate the spring; and a holding bar, which holds the hammer down till released by the catch. Every one of these parts is indispensable for the mouse trap to function. Without one of these parts, the trap would be useless. Essential is that all parts have been put together, have been designed, with an *intention*: to kill mice. The coming together of the parts at the same time and in the right way cannot

---

41. Daniel Dennett: *Darwin's Dangerous Idea*, pp. 251, 249 (italics in the text).
42. Michael Behe: op. cit., p. 39.

be explained as having taken place by chance, one independently after the other, and with time intervals.

Nature consists entirely of irreducibly complex systems, from the ones measuring microns, like the flagellum of a bacterium, to systems active over part or the whole of an organism, like the immune system. From one angle the systems of an organism are ordered from the whole down to its organs, cells, molecules, atoms and subatomic particles; from the opposite angle the order climbs from the elementary particles through the cells and the organs to the whole organism, be it plant or animal. "For instance, the emergence of a sensory organ like the eye, even the most elementary [in lower animals], requires the appearance in a coordinated manner of a whole sequence of transformations. It is difficult to understand how each single mutation might be selected before the eye is functional, as it is its function that is the necessary criterion of the selection."[43] Richard Lewontin concludes: "It is not that the whole is more than the sum of its parts. It is that the properties of the parts cannot be understood except in their context in the whole."[44]

The appearance of having been designed is undeniable to anyone contemplating the intricacy and efficacy of nature, the 'wonders' of nature, even to Richard Dawkins, for he writes in the first page of *The Blind Watchmaker*: "Biology is the study of complicated things that give the appearance of having been designed for a purpose." Yet being the arch-Darwinian whom he declares himself to be, he has to deny any acceptance of design, for that was precisely what Darwin erected his theory against. Still, "Dawkins doesn't just grudgingly acknowledge some faint impression of design in life; he insists that the appearance of design, which he ascribes to natural selection, is overpowering," comments Behe. He goes on to quote Dawkins: "Yet the living results of natural selection overwhelmingly impress us with the appearance of design as if by a master watchmaker, impress us with the illusion of design and planning."[45]

Intelligent Design, instead of systematically rejecting the logical conclusion of scientific findings for dogmatic reasons, accepts it. Irreducibly complex systems clash with Darwinian theory because they cannot be explained in the gradual, step-by-step manner envisioned by Darwin, who wrote: "If it could be demonstrated that any complex organ existed, which could not possibly have been formed by numerous, successive,

---

43. Gerard Amzallag: *L'Homme végétal*, p. 58.
44. Richard Lewontin: *The Doctrine of DNA*, p. 122.
45. Michael Behe: op. cit., p. 264.

slight modifications, my theory would absolutely break down."[46] Intelligent Design holds this to be demonstrated, therefore steps outside the Church of Darwinism and becomes anathema to it.

"It is important to understand that intelligent design is not a claim that miracles occur. Rather, (1) design is an actual feature of the universe, one that cannot be duplicated by the effects of natural law and chance, and (2) design is researchable, and therefore a valid part of science." But: "The question is not whether the universe shows evidence of design. Of course it does. The question is how to interpret the evidence." (O'Leary) In the words of William Dembski, who considers himself a neutral proponent of ID: "Intelligent design is not a form of anti-evolutionism. Intelligent design does not claim that living things came together suddenly in their present form through the efforts of a supernatural creator. Intelligent design is not and never will be a doctrine of creation."[47]

Problems of Intelligent Design

Neutral ID theory is struggling with problems of which it does not seem to be fully aware. The first one is that, while doing its utmost to comply with the rules of modern science, it does not even consider the gradations of reality traditionally called the 'Chain of Being'. It cannot disregard them, for after all the 'design' is about the phenomena of life, and the 'intelligence' is about mind which is consciousness. One finds evidence of this anomaly for instance in Michael Behe's recent reflections on his own view: "Such a view implies ultimately that life is an intrinsic property of matter, that the course of evolution is directed by natural law, and that our own existence was ordained in nature from the beginning. ... Such a view is intellectually exciting because it holds out the prospect of a final union of biology and physics, and thus of a fully rational and lawful biology."[48] Life an intrinsic property of matter? A final union of biology and physics?

A second point at issue is that Intelligent Design is not a theory, but a question, a problem for which 'Darwinism' and other materialistic evolutionary hypotheses have no solution. The facts which led to the question are indubitably scientific and the deductions logical. They

---

46. Charles Darwin: *The Origin of Species*, p. 171.
47. Denyse O'Leary: op. cit., pp. 177, 243..
48. William Dembski (ed.): *Uncommon Dissent*, p. 176.

are that the irreducible or specified complexity omnipresent in nature cannot be explained by step-by-step gradualism, because in any given natural 'mechanism' or structure each part must be functional simultaneously with all others to render the 'mechanism' or structure efficient. This requires an intention, a plan, a blueprint – a design, which means that in nature a designing intelligence is at work.

This reconsideration or reintroduction of consciousness by scientists in biological matters, together with the anthropic principle in physics, may be of decisive importance for the future of science. However, there is as yet no theoretical scaffold to support the scientific statement of ID and the question it asks from academic science. What is more, there cannot be a theory as long as ID, besides accepting an intelligent agent of the design, remains stuck in matter. The paradigm shift has to be considerable, but it is imperative.

A third problem is that the ID movement as a whole seems to be little aware of its limitations to its Western, Christian background. As such it is part of the far from extinct 'Eurocentric' attitude of a race which conquered the globe with its technology, commercialism, Christian religion, and true God. The spirituality of the inner explorations by its own great souls and mystics it has suppressed or ridiculed; the great spiritual movements, especially those in the East, it has covered with disdain. "In the Western world theology is wedded to a particular concept of God." (O'Leary) "We can take the term 'purposeful designer' in a very broad sense to refer to any being, principle or mechanism *external* to our universe." (Nick Bostrom) For the Judeo-Christian God thrones above his handicraft, and the idea that God might be the *internal* sustainer of life's evolution on Earth is usually condemned as pantheism. Once again the words of the physicist David Bohm, who had the courage to explore beyond the invisible but still very real conceptual Western boundaries, deserve our attention: "Ultimately the origin of all this lies in the creative intelligence, which is beyond anything that can be discussed in the manifest physical side. This intelligence is universal and acts in every area of mental operation."[49]

Inevitably, Intelligent Design has been and is severely criticized. "Intelligent Design theory is pernicious nonsense which needs to be neutralized." (Nial Shanks) "Intelligent design is a hopeless theory. ... It makes no successful predictions, it fails to unify diverse classes of phenomena, and it has garnered no support for the alleged character

---

49. David Bohm and David Peat: op. cit., p. 219.

and abilities of the designing agent or agents. It is on a par with the hypothesis of disco-dancing fairies." (Tim Lewens) "Any creative intelligence, of sufficient complexity to design anything, comes into existence only as the end product of an extended process of gradual evolution. Creative intelligences, being evolved, necessarily arrive late in the universe, and therefore cannot be responsible for designing it. God is a delusion, and a pernicious delusion," writes the author of *The God Delusion*, Richard Dawkins.[50]

Nonetheless, Intelligent Design is gaining ground, despite the damage done to its genuine scientific position by interested religious forces. Creationist minds know how to gild their case with apparently neutral arguments. Poorly informed media contribute to the confusion about this potentially crucial event in the biological sciences.

The following is the conclusion of Jean Swyngedauw, a French physician: "It is not biochemistry which, in the course of time, has guided the meandering lines of the evolution. Biochemistry might have taken an unspecified number of directions or not realized anything at all. Constant information has been necessary, first for the multiplication of the cell to become possible, then for the genealogical tree of the species to spread out as reconstructed by the paleontologists. ... There is no alternative: constant information must have borne the essential responsibility of the evolution. ... Information has been indispensable, but what is its source? We are impelled to attribute life to a kind of infiltration of the spirit in the domain of matter-energy. ... Present everywhere and not bound to time, its nature is perforce metaphysical."[51]

---

50. Richard Dawkins: *The God Delusion*, p. 31.
51. Jean Swyngedauw: *À l'origine de la vie le hasard?* pp. 160-62.

15.

Intelligence that is Consciousness that is Being

Neither the laws nor the possibilities of physical Nature can be entirely known unless we know also the laws and possibilities of supraphysical Nature.
This is what we call evolution which is an evolution of Consciousness and an evolution of the Spirit in things and only outwardly an evolution of species.

SRI AUROBINDO

West and East

Science, a search for objective truth, is in every one of its aspects and practices intertwined with the cultural and religious ideas of its time – as we have seen throughout our exploration of evolution and the evolutionary theories. We have also found that modern science, a product of the Western mind, reasons almost exclusively within the framework of the main Western religion, Christianity. This defines its references to history and scripture, and its concept of God. Might there then be something worthwhile Eastern spirituality has to tell about modern science in general and evolution in particular?

In 1949, Sri Aurobindo wrote in a message: "East and West have the same human nature, a common human destiny, the same aspiration after a greater perfection, the same seeking after something higher than itself, something towards which inwardly and even outwardly we move. There has been a tendency in some minds to dwell on the spirituality or mysticism of the East and the materialism of the West; but the West has had no less than the East its spiritual seekings and, though not in such profusion, its saints and sages and mystics; the East has had its materialistic tendencies, its material splendours, its similar or identical dealings with life and Matter and the world in which we live. East and West have always met and mixed more or less closely, they have powerfully influenced each other and at the present day are under and increasing compulsion of Nature and Fate to do so more than ever before."[1]

The outward Western and the Eastern approach of the underlying Truth differ considerably. These too are Kuhnian paradigms, centuries old, of which the tenets have hardened to the point of apparently becoming irreconcilable. While theoretical physics is regarded with awe by the uninitiated, all matters psychological, philosophical and religious in West and East are supposed to be accessible to anyone. All the same: "It is not every untrained mind that can follow the mathematics of relativity or other difficult scientific truths or judge of the validity either of their result or their process. All reality, all experience must indeed, to be held as true, be capable of verification by a same or similar experience, so, in fact, all men can have a spiritual experience and can follow it out and verify it in themselves, but only when they have acquired the capacity or can follow the inner methods by which that experience and verification are made possible." (Sri Aurobindo, LD 650-51[2])

Given the millennia of its history, the riches of its secular and spiritual literature, and the exceptional beings representing great peoples and large spiritual movements, a closer acquaintance with the unfamiliar East should at least be worth our interest. But true understanding always requires a long and assiduous effort, something the impatient West is rarely willing to make. Yet, what particular interests are resisting to provide, the trend of history seems to be bringing about in the rapid integration of humanity.

The basic concept behind all worldviews, affirmations or denials, is always the concept of the Fundamental, or Ultimate, or Essential, or Divine, or That: 'God'. In the Vedanta this is called "Brahman,"[3] the Absolute, the One. "Brahman is the Alpha and the Omega. Brahman is the One besides whom there is nothing else existent." (LD 33) For "Brahman is in all things, all things are in Brahman, all things are Brahman." (LD 139) As an Upanishad says: "O Brahman, thou art this old man and boy and girl, this bird, this insect." (LD 328)

The One as omnipresent Reality crucially differs from the Western concept of God. The One is *in* this old man, this rock, this raindrop, while Jahweh, and consequently the Christian God, resides above,

---

1. Sri Aurobindo: *Autobiographical Notes*, p. 551.
2. Sri Aurobindo's works will be referred to in the text by the following abbreviations: EDH: *Essays Divine and Human;* EPY: *Essays in Philosophy and Yoga;* HC: *The Human Cycle;* LD: *The Life Divine;* LY: *Letters on Yoga;* Sav: *Savitri;* SY: *The Synthesis of Yoga.*
3. Not to be confused with 'Brahma', the creator god of the Hindu trinity.

outside his creation. Christian theology condemns the inherence of the One in his manifestation as pantheism. So it has done with stoicism and Spinozism, and so it does with what is vaguely named 'orientalism'. What it does not realize is that the One, Brahman, is more complex than this: there is the silent or passive Brahman ('the Ineffable') and the active, manifesting Brahman. "In fact, the Brahman is one not only in a featureless oneness beyond all relation, but in the very multiplicity of the cosmic existence." (LD 641) From this follows that the One does not 'create': as there is nothing but itself, it can only manifest itself from itself. (Contemporary gender sensitivity together with the line of this argument necessarily lead in this case to the use of the neutral 'it'.) All is That. Evolution is an unfolding *in* the One.

The view that God is in each being, and therefore in the heart of each human being, is not foreign to the West, but has been ruthlessly suppressed by the religious institutions which held that God could only be approached through them. Direct personal experience of God was forbidden. 'Mystics', people who had direct experiences and confessed they had them, were persecuted, even those later canonized as saints, and many were cruelly executed as heretics. Therefore all "God within" movements had to go underground, as one finds in the fascinating history of Gnosticism, spiritual alchemy, Hermetism, Rosicrucianism, and Western spiritual occultism in general.

In the beginning of the first chapter we have already learned that evolution is inherent in the texts of the ancient Seers, because it is how things actually are. "In certain aspects the old Vedantic thinkers anticipate us," noted Sri Aurobindo, "they agree with all that is essential in our modern ideas of evolution. ... The Puranas admit the creation of animal forms before the appearance of man and in the symbol of the Ten Avatars trace the growth of our evolution from the fish through the animal, the man-animal and the developed human being to the different stages of our present incomplete evolution." (EDH 385)

The "avatar" represents a divine intervention in the earthly evolution when it has reached a stage where, without such an intervention, it cannot go further. Statues of these avatars, pictured symbolically at evolutionary key periods, are found everywhere in the temples of India and of countries where Hinduism spread in the past: Fish (eight ninths of evolution took place in water), Tortoise (amphibian), Boar (mammal), Dwarf (hominid), Rama with the ax (*Homo habilis* or *ergaster*), Rama with the bow (*Homo sapiens*), followed by the spiritual evolution of humanity in Krishna, the Buddha, and Kalki, the future avatar.

This "procession of the avatars" is unmistakable. It should be clear that "avatar" in this context has no relation to the present degradation of this concept in some sectarian movements. It is also noteworthy that the Hindu concept of the avatar is applicable to the person and incarnation of Jesus the Christ on Earth, the one and only, while the Hindu series of avatars sustains the evolution of consciousness in its increasing complexity.

"Once the evolutionary hypothesis is put forward and the facts supporting it are marshaled, this aspect of the terrestrial existence becomes so striking as to appear indisputable," (LD 836) wrote Sri Aurobindo (1872-1950) in *The Life Divine*, the first and major formulation of his evolutionary vision. Trained as a classical scholar at Cambridge, he also acquired an encompassing command of the rich literature of India in her various languages, and penetrated ever deeper in her ancient scriptures as he advanced in his own spiritual experience. A radical in spirituality as he was in the liberation politics of his country, he concluded that the fundamental truth of Veda and Vedanta had to be accepted and put into practice unconditionally. If all is That, if all is the One, the logical consequence of this premise is tremendous: matter too is the One, the Earth and everything on it is the One; life on Earth is not an error or an illusion, it has its meaning in the One because the One *is* meaning. The aim of true spirituality, therefore, should not be an escape to a Hereafter or Nirvana, it should be the realization of the earthly potential, the transformation of our earthly existence – practically speaking the realization of the next step in evolution.

The magnitude of this logical conclusion has been little realized until now, and still less put into practice. Rooted in the Hindu spiritual tradition, it nevertheless confronts head on some of its major movements, like the illusionist teachings of Shankara and the Buddha. Matter, our own substance, is like everything else the One. "Matter itself is substance and power of spirit and could not exist if it were anything else, for nothing can exist which is not substance and power of Brahman ... Matter is a form of Spirit, a habitation of Spirit, and here in Matter itself there can be realization of Spirit." (LD 761, 665) It should therefore be the aim of spirituality not to try to escape by the shortest way from this earthly life into everlasting ecstasy or nonexistence, but to grow into the One and collaborate with its intentions in this material realm of its manifestation. This must be the reason our souls have incarnated here, and if they are "a spark of the Divine" it must be the rationale of their incarnation.

Sri Aurobindo was not only a classical scholar and a generally recognized master of the English language, he was also widely read in cultural, political and scientific matters. His evaluation of science was not the negation by the stereotypical bearded mystic, it was positive. Behind the mono-dimensional view of scientific materialism he saw the contribution of science to the integration of humanity, and its enlarging of the knowledge and conditions of the physical realm necessary for the evolutionary step forward which he saw as imminent. "For that vast field of evidence and experience which now begins to reopen its gates to us, can only be safely entered when the intellect has been severely trained to a clean austerity ... It became necessary for a time to make a clean sweep at once of the truth and its disguise in order that the road might be clear for a new departure and a surer advance. The rationalistic tendency of materialism has done mankind this great service." (LD 11)

Involution and Evolution

Ignoring a problem does not solve it. How did life appear in matter? How does an organism consisting only of matter produce consciousness? The mind-body problem remains vividly discussed though unexplained, and the present gross materialistic stance of neurobiology, that all is matter and that consequently life and consciousness are (epi)phenomena of matter, raises more questions than it answers. A television set is made of material components, yet nobody will contend that the news reader, the ice skater or the soap opera on the screen is produced by the material components of the set.

"We do not see or know, but it is expounded to us [by materialistic science] as a cogent account of Nature-process, that a play of electrons, of atoms and their resultant molecules, of cells, glands, chemical secretions and physiological processes manages by their activity on the nerves and brain of a Shakespeare or a Plato to produce or could be perhaps the dynamic occasion for the production of a *Hamlet* or a *Symposium* or a *Republic*; but we fail to discover or appreciate how such material movements could have composed or necessitated the composition of these highest points of thought and literature ... These formulae of Science may be pragmatically correct and infallible, they may govern the practical how of Nature's processes, but they do not disclose the intrinsic how or why; rather they have the air of formulae of a cosmic Magician, precise, irresistible, automatically successful each in its field, but their rationale is fundamentally unintelligible." (Sri Aurobindo, LD 299)

One explanation is the following. "Nothing can evolve out of Matter which is not therein already detained." (LD 87) Put in the broader context: evolution can only be the result of a previous "involution", "evolution is an inverse action of the involution." (LD 853) "In a sense, the whole of creation may be said to be a movement between two involutions, Spirit in which all is involved and out of which all evolves downward to the other pole of Matter, Matter in which all is involved and out of which all evolves upwards to the other pole of Spirit." (LD 129) (It may be noted that Sri Aurobindo sometimes uses the word 'creation' where 'manifestation' is meant, as nothing can be created anew from outside the One – there is no 'nothing' – and all that exists is manifested by the One out of itself.)

The universal scheme behind the terrestrial evolution may be drafted as follows. There is the silent Brahman, the Ineffable, which can only be known by the highest spiritual experience. One and the same is the active Brahman which manifests its inexhaustible infinity in endless time and timelessness. Manifestation has been and will be always. At the top of the gradations of the eternal manifestation are the worlds of the three ultimate attributes of the One: *sat* (existence), *chit* (consciousness), *ananda* (often translated as bliss, the quintessence of supreme ecstasy). The fourth domain of this upper hemisphere has been called "Supermind" by Sri Aurobindo. It is the manifesting consciousness of the One, the unity-consciousness or truth-consciousness, the knowledge that is truth that is power.

This Supermind is not something like the human mind at its highest potential. It is a supra-mind from which our human mind derives, but which is as different from it as the sunlight is from the ray of a distant star. It is what some scientists have intuited behind the complexities of nature, what is present in nature and supports it everywhere. It is the real "mind of God." In fact, it is what is called 'God' in the purest sense.

From the domains of the Supermind descend the worlds of the lower hemisphere, those with which we are more familiar: mind, life, matter. Seen as such, the One has projected itself into its contrary, which means that below matter – already an astonishing degree of organization as modern science has found out – there are still deeper, darker levels, those of the subconscient and of the inconscient foundation. Nevertheless, all is and remains the One in every differentiation: "Matter too is the Brahman," and so are life and mind. "All is in each as well as each

264

in all." (LD 21) "The One is the fundamental truth of existence." (LD 358) "All is fundamentally the same substance, the same consciousness, the same force, but in different forms and powers and degrees of itself." (LD 120)

There are countless worlds on every level of the manifestation, from the highest to the lowest, from *sat, chit, ananda* and the brilliances of the Supermind to the worlds of mind and life, higher and lower. There is "this world that we see and those other worlds that we do not see." (LD 42) "You speak as if the evolution were the sole creation," Sri Aurobindo wrote in a letter. "The creation or manifestation is very vast and contains many planes and worlds that existed before the evolution, all different in character and with different kinds of being. The fact of being prior to the evolution does not make them undifferentiated." (LY 385) Yet he called those non-material worlds "typal," meaning that their beings – gods, angels, demons big and small – are not subjected to change, but that they are perfect and perfectly happy in their state, however evaluated by us.[4]

Ours is not a typal but an evolutionary world. As all possibilities exist in the One, there was, and always is, also the possibility of taking on the appearance of its contrary: infinite oneness becomes infinite fragmentation; light becomes darkness, truth falsehood, bliss suffering, eternal existence death. While embodied in matter, this "appearance" may seem rather harsh to us, but the One remains always one, even when apparently split up into the infinitesimal. This adventure of the One to rediscover and recover itself explains the evolution as an increase in consciousness, which on the human level has become self-conscious, but which is on the whole gamut of existence only somewhere halfway.

This view is as vast as it is concrete and detailed. "The emergence of the movement from the Immutable [the One] is an eternal phenomenon and it is only because we cannot conceive it in that beginningless, endless, ever new-moment which is the eternity of the Timeless that our notions and conceptions are compelled to place it in a temporal eternity of successive duration to which are attached the ideas of an always recurrent beginning, middle and end. ... The universe persists or always comes back into manifestation because the will to become is eternal and must be so since it is the inherent will of an eternal Existence." (LD 76, 669)

There is, on the other hand, the full awareness of the apparent

---

4. This "multiverse" has been evoked by Sri Aurobindo in his epic poem *Savitri*.

insignificance of planet Earth in the universe, repeated with spite time and again in the writings of scientific materialists, and called "the Copernican Principle." Our Earth, "this minute island of life," spins "in a small corner of the universe ... inconspicuous among the immensities." Still, this appraisal is the result of "the illusion of size, of quantity, that induces us to look on the one as great, the other as petty." (LD 73) For the Earth, "a habitable planet in an uninhabitable system," is connected to the other worlds, and those other worlds have access here.

"The experiment of human life on an earth is not now for the first time enacted," wrote Sri Aurobindo in one of his old and now deciphered notebooks. "It has been conducted a million times before and the long drama will again a million times be repeated. In all that we do now, our dreams, our discoveries, our swift or difficult attainments we profit subconsciously by the experience of innumerable precursors and our labour will be fecund in planets unknown to us and in worlds yet uncreated. The plan, the peripeties, the denouement differ continually, yet are always governed by the conventions of an eternal Art. God, Man, Nature are the three perpetual symbols. (EDH 141)

The gradations, planes or levels of being constitute the backbone of all schools of wisdom since ancient times. In the Hindu scriptures the worlds of matter (or substance), life and mind are known as "the three strides of Vishnu." "Matter is only energy in action, and as we know in India, energy is force of consciousness in action." (LY 222) "There is no such thing as the self-evident Matter posited by nineteenth century Science," (LY 224) leading to the circle of metaphysical materialism, its magic formula, that all is matter because there is nothing but matter.

If such is Reality, then a time will come that science, physical and biological, will have to confront and accept it. No doubt, Galileo's method, developed into what became called the scientific method, was a necessary start to get a grip on the aspect of reality perceptible by the senses. But it is now clear to open-minded theorists that a new and huge step will have to be taken. The indications of this necessity are many and of critical importance: the unsolved incompatibility between relativity and the quantum theories; the conflict between the Einsteinian and Bohrian concepts of reality; the public statement by leading physicists that physics is in crisis; the new discoveries in genetics which call 'Darwinism' into question ...

"The material interpretation of existence was the result of an exclusive concentration, a preoccupation with one movement of Existence, and such an exclusive concentration has it utility and is therefore permissible; in recent times it has justified itself by the many immense and the innumerable minute discoveries of physical Science. But a solution of the whole problem of existence cannot be based on an exclusive one-sided knowledge; we must know not only what Matter is and what are its processes, but what mind and life are and what are their processes. And one must know also spirit and soul and all that is behind the material surface: only then can we have a knowledge sufficiently integral for a solution of the problem." (LD 652)

In the course of our exploration, we have seen on several occasions how the gradations of the Chain of Being influenced instinctively the thought of even some of the most hardened materialistic thinkers and scientists, e.g. Bertrand Russell and Jacques Monod. The reason is that reality exceeds the artificial separations constructed by the mind, not vice versa, and that the workings of reality, or nature, cannot be overlooked permanently. By way of illustration: "I suppose a matter-of-fact observer, if there had been one at the time of the unrelieved reign of inanimate Matter in the earth's beginning, would have criticized any promise of the emergence of life in a world of dead earth and rock and mineral as an absurdity and a chimera; so too, afterwards he would have repeated this mistake and regarded the emergence of thought and reason in an animal world as an absurdity and a chimera. It is the same now with the appearance of supermind in the stumbling mentality of the world of human consciousness and its reasoning ignorance." (LY 8-9) For if "all evolution is in essence a heightening of the force of consciousness in the manifest being" (LD 720), why should it stop at humanity?

That matter is energy had become a supposedly understood platitude till the atomic bombs showed what the equation actually meant. To the acceptance and understanding of the equation matter=energy=consciousness, the mind of humanity has not yet arrived, although it is one of its oldest occult truths. "The inconscience of Matter is itself a hooded consciousness." (EDH 165) "Matter is a blind form of the Spirit." (EPY 571) If material energy is that powerful, how powerful is the power of consciousness which it contains? Powerful enough to create quasars, galaxies and multiple universes, and to perform the wonders of the subatomic world which can be represented schematically, but which surpass the imagination.

"The material universe is only the façade of an immense building which has other structures behind it, and it is only if one knows the whole that one can have some knowledge of the truth of the material universe. There are vital, mental and spiritual ranges behind which give the material its significance." (LY 212) "Earth is the foundation and all the [typal] worlds are on the earth, and to imagine a clean-cut or irreconcilable difference between them is ignorance; here and not elsewhere, not by going to some other world, the divine realization must come." (LY 178) But this is another matter.

Mother Nature

The ancient cultures felt nature as a bountiful Mother bringing forth and caring for the world and the countless things and beings in it. However, this Mother, good and generous on most occasions, could also be severe in her actions. For one of her many aspects was the Queen of the Universe laying down its laws – *Maya* in India, *Maat in Egypt* – and on occasion apparently overruling them all as Fate. Behind her decisions were the cosmic objectives, unfathomable by the human mind.

A time came that science was given to the human race, which felt that knowledge would mean power. For his knowledge to be effective, the scientist had to be convinced that his intellect was the supreme instrument by which he could formulate the laws of the universe. The combined effort of scientists and technicians, to which many of the best minds dedicated their lives, has worked wonders: in a short time it has changed the Earth, in the process making humanity one. But of Mother Nature they have explored only the most superficial aspect, the material one. Now they are increasingly feeling restricted by the mental walls within which they have immured themselves. Breaking them down may be a painful enterprise, but it will be necessary.

"I mean by Nature only the aggregate action and product of many natural laws, and by laws the sequence of events as ascertained by us," wrote Charles Darwin in *The Origin of Species*.[5] The word "only" in this sentence tells us that he wanted to transfer nature, as the whole of life's evolution on Earth, within the rational domain of scientific materialism from that of divine Providence, the Great Mother of the religion of his culture. "Natural laws" should determine the precise way in which mechanisms or processes function, preferably in mathematical formulae, although Darwin would have been satisfied with "ascertaining sequences of events".

---

5. Charles Darwin: *The Origin of Species*, p. 88.

He too dedicated his life to science, to finding out the ways of "transmutationism", the processes of evolution. He guessed or hypothesized, everything duly considered, that new species would appear through the gradual accumulation of minute changes, adaptation, and natural selection. Constant research has discovered many wonderful mechanisms and processes of life; diligent searching has revealed a vertiginous diversity of life in the Earth's past, as well as dramatic changes in the planet itself. The sciences utilized for the study of evolution have accumulated a wealth of data suggesting "sequences of events". But theories keep cropping up or reappearing, and rock-solid 'laws', in the strict sense of the word, have not been established.

Scientific theories depend on the understanding of reality, in this context another word for nature. If reality consists of the Chain of Being, then science, as we have seen, must take into account the levels other than matter. If all is the One, then nature in her entirety is the One. "What Nature does is in reality done by the Spirit." (LD 355) "All her works are instinct with an absolute intelligence," (SY 91) which is a supra-rational, supra-mental intelligence conscious of the whole of Existence in a simultaneous perception within and without Time. In the One, in Brahman, there cannot be anything that is not the One or Brahman – this is the supreme monotheism – nor can there be any 'error', for then the One would not be perfect. This leads to the logical conclusion: "When we speak indeed of the errors of Nature, we use a figure illegitimately borrowed from our human psychology and experience, for in Nature there are no errors but only the deliberate measure of her paces traced and retraced in a prefigured rhythm, of which each step has a meaning and its place in the action and reaction of her gradual advance." (IHU 99)

"If we look carefully at these workings of Nature, once we put aside the veil of familiarity and our unthinking acquiescence in the process of things as natural because so they always happen, we discover that all she does in whole or in parts is a miracle, an act of some incomprehensible magic. The being of the Self-existence and the world that has appeared in it are, each of them and both together, a suprarational mystery. There seems to us to be a reason in things because the processes of the physical finite are consistent to our view and their law determinable, but this reason in things, when closely examined, seems to stumble at every moment against the irrational or infrarational and the suprarational: the consistency, the determinability of process seems to lessen rather than increase as we pass from matter to life and from life to mentality;

if the finite consents to some extent to look as if it were rational, the infinitesimal refuses to be bound by the same laws and the infinite is unseizable. As for the action of the universe and its significance, it escapes us altogether ..." (LD 326-27)

Words like 'magic', 'suprarational' and 'infrarational' may easily evoke the misty 'mysticism' so scathingly derided by the positivist mind. But this would be in direct contradiction with the accuracy the humans and their world have been explored for centuries by the Western and especially the Eastern spiritual genius, of which modern science has no knowledge. In the introduction to *Le Passage de la matière à la vie selon le Bouddha Gauthama* (the transition from matter to life according to the Buddha Gauthama) by Emmy Guittès, the publisher writes: "This little book is a real revelation, for it changes everything one thought one knew about the ignorance of the past centuries! The composition of the atom and the explosive effects of its splitting, the relativity of time, the expansion of the universe, an energy which is in constant transformation but never exhausted, an incalculable number of min-iature universes repeated infinitely, life and death of the stars, and many other scientific processes have been expounded by Siddhartha Gauthama, the Buddha, and included in the voluminous *Abbidhama*."[6] From Sri Aurobindo's writings we choose three topics related to science.

"We can say that the Big Bang theory is currently regarded as a well-established theory, the standard-model acceptable to most physicists, and that the questions that remain do not cast serious suspicions on it," wrote Kitty Ferguson in 1994.[7] In 1933, however, the universe was still assumed to be eternal and unchanging, and the Big Bang model, endorsed by Einstein, blatant nonsense. Around 1950 the controversy raged between Hoyle, Bondi and Gold's Steady State model of the universe and the theory of the explosion at the beginning of time. Sri Aurobindo finished his epic *Savitri* shortly before his passing in 1950. Aware of the competing trends in science, he wrote nonetheless:

A Mystery's process is the universe.
At first was laid a strange and anomalous base,
A void, a cipher of some secret Whole,
Where zero held infinity in its sum
And All and Nothing were a single term ...

---

6. Emmy Guittès: *Le passage de la matière à la vie*, p. 9.
7. Kitty Ferguson: *The Fire in the Equations*, p. 126.

A slow reversal's movement then took place:
A gas belched out from some invisible Fire,
Of its dense rings were formed these million stars … (Sav 100-01)

Decades before he had already noted, interpreting the old scriptures: "As in the immobile ether arises, first sign of the creative impulse of Nature, vibration, *Shabda*, and this vibration is a line of etheric movement, is ether contacting ether in its own field of mobile self-force and that primal stir is sufficient to initiate all forms and forces, even such is the original movement of the Infinite." (EDH 198) The language is quite different from the modern scientific terminology (to Sri Aurobindo 'ether' was 'space'). But the point is that what at the time was a controversial novel theory is stated as fact in a *magnum opus* intended to contain his spiritual heritage. Moreover, terms like "anomalous base" and "the etheric movement of vibration" bring to mind the quantum theory according to which virtual particles arise continuously out of "the void" and may be at the origin of the universe.

Another topic is the ancient concept of *pralaya*, the cyclic origin and extinction of the universe. This was even recently thought to belong to the old Hindu lore, just like the cycles of time were part of the mythology of ancient Greece. But see, not only is the universe now supposed to have evolved from a magic primeval particle, the question "what came before the Big Bang" has become scientifically legitimate, as has the question what to expect after the Big Crunch, when our universe dies. Paul Steinhardt and Neil Turok have published a book with the title *Endless Universe: Beyond the Big Bang.* "They say they were motivated to form a new theory as the big bang came to require more and more exotic elements – inflation, dark matter, dark energy – to make it fit observations."[8] And string theory has spawned not only the possibility of a multitude of universes, a 'multiverse', but an infinity of them.

Paul Steinhardt writes: "In the cyclic model, the Big Bang is not the beginning but, rather, an event that has been repeating every trillion years, extending far into the past. Borrowing ideas from string theory, the cyclic model proposes that each bang is a collision between our tridimensional world and another three-dimensional world along an extra spatial dimension. Each bang creates new hot matter and radiation that begins a new period of expansion, cooling, galaxy formation, and life,

---

8. *Scientific American India*, September 2007, p. 80.

but space and time exist before and after the bang."[9] How shall the intellectual outsider choose between this mathematical hypothesis and the ancient wisdom – which is only wisdom if it is knowledge, though not necessarily mental knowledge? "What will happen," asks Claude Allègre, "if it is proven tomorrow that the big bang does not define the beginning of everything, but only the beginning of one episode among the innumerable which the universe has known, and if one proves that, finally, the history of the cosmos is nothing but alternating periods of expansion and contraction?"[10]

A third topic could be called *accelerando e crescendo*, the acceleration and intensification of the cosmic evolution. In *The Life Divine* (LD 932), Sri Aurobindo wrote: "The first obscure material movement of the evolutionary Force is marked by an aeonic graduality." This is the evolution of the material universe from the Big Bang to the formation of the solar system and planet Earth. "The movement of life-progress proceeds slowly but still with a quicker step." This is the evolution of life on Earth from its origin, now estimated at about 4 billion years ago, till the appearance 2 million years ago of our own species, now spread in great numbers in all habitable parts of the planet. "Mind can still further compress the tardy leisureliness of Time and make long paces of the centuries." The acceleration and intensification of mind in science and technology has presently reached such a momentum that it causes a global sense of insecurity, not to say outright fear for tomorrow. The curve of this acceleration and intensification is an undeniable illustration that the universe, with humanity as a decisive element, is reaching a point where something drastic has to happen, if it is not already happening.

"In a seminal academic paper delivered to a NASA colloquium Vernor Vinge wrote: 'I argue in this paper that we are on the edge of change comparable to the rise of human life on Earth.' He is anticipating the possibility of greater-than-human intelligence. He is talking about some form of transcendence. ... As a metaphor for mind-boggling social change, the 'singularity' has been borrowed from math and physics. In those realms, singularities are the points where everything stops making sense.[11] ... Some people think we are approaching such a Singularity

---

9. John Brockman (ed.): *What Are You Optimistic About?* p. 64.
10. Claude Allègre: *Dieu face à la science*, p. 94.
11. The common example of a singularity is the explosion of the primeval particle which gave birth to the universe. At this 'magic' moment there were no physical laws and therefore no explanations. Physicists accept this miracle because they reason

– a point where our everyday world stops making sense. They think that is what happens when The Curve goes almost straight up. ... He believes some sort of fundamental transcendence will happen soon."[12] In Sri Aurobindo's language this reads: "When the conscious Spirit intervenes, a supremely concentrated pace of evolutionary swiftness becomes possible." (LD 932)

Evolution Two-Tiered

"We can no longer suppose that God or some Demiurge[13] has manufactured each genus and species ready-made in body and in consciousness and left the matter there, having looked upon his work and seen that it was good," wrote Sri Aurobindo. "It has become evident that a secretly conscious or an inconscient Energy of creation has effected the transition by swift or slow degrees, by whatever means, devices, biological, physical or psychological machinery – perhaps having made it, did not care to preserve as distinct forms what were only stepping stones and had no longer any function nor served any purpose in evolutionary Nature. But this explanation of the gaps[14] is little more than a hypothesis which as yet we cannot sufficiently substantiate. It is probable at any rate that the reason for these radical differences is to be found in the working of the inner Force and not in the outer process of the evolutionary transition ..." (LD 709-10)

In this passage Sri Aurobindo avows the difference between the outer, materialistic (e.g. 'Darwinian') explanation of evolution, and its explanation (e.g. his own) through an inner Force. Yet, he also points out that ascribing the works of nature to "an inconscient Energy of creation" – or to an Intelligence as in the theory of intelligent design – is a hypothesis which has not been sufficiently substantiated. Let us see how he himself accounts for the workings of evolution.

We have now a rudimentary idea of the general framework of the universal manifestation. The ineffable, passive One remains eternally

---

backwards from their explanation – the 'laws' – of the universe as perceived now by their physical senses and their instrumental extensions.

12. Joel Garreau: *Radical Revolution*, pp. 71-1.

13. A "Demiurge" is a secondary God who has created the world of his own volition, and is responsible for the falsehood, suffering and evil in it. In Gnosticism he was identified with Jahweh.

14. "The God of the gaps" is an ironic phrase for the fictitious explanations of gaps in the understanding of scientific matters.

absorbed in itself; the active One, which is one and the same, manifests itself in a scale of the gradations of existence, from the highest to the lowest – to the apparent opposite of its essential attributes. Every gradation of this involution represents a 'multiverse' of typal worlds and beings. "The creation has descended all the degrees of being from the Supermind to Matter and in each degree it has created a world, reign, plane or order proper to that degree." (LY 1)

Our world emerges from the densest point of the One's involution into its contraries; it evolves from the Negation, the utter Inconscient, and climbs up step after previously established step, thus integrating the aspects of the One extant in eternity. "Before there could be any evolution, there must needs be this involution of the Divine All that is to emerge. Otherwise there would have been not an ordered and significant evolution, but a successive creation of things unforseeable, not contained in their antecedents, not their inevitable consequences or right followers in sequence." (EDH 162)

On our Earth, the gradations that have been integrated until now are the three lower ones: matter, life, mind. Our eyes are only able to perceive what we call 'matter', but this Earth, and the life and mental consciousness on it, remain connected to their involutionary planes of origin. "The immense material world in which we live is not the sole reality but only one of innumerable potential and existent universes; all of them need not have either Matter as we know it or the Inconscient as we know it for their base. Indeed, this world of Matter is itself dependent on many planes of consciousness and existence which are not material; for these have not this gross substance as their foundation or as the medium of their instrumentation of energy and consciousness or their primary condition of existence." (EDH 241)

Crucial to the understanding of this view is that the One (for reasons of its own which can only be *ananda* or delight) has chosen to plunge into its opposite which we call the Inconscient. But the One can never be two, consequently the Inconscient is also the One, and all the attributes of the One are contained, though hidden, in the Inconscient. The first manifestation out of the Inconscient is Matter, already a high degree of organization (as particle physics has discovered). And it is on this basis of Matter that the gradations of the involution are recovered and one by one built up again, up to their Origin.

All the levels of being are aspects of the One and therefore interdependent. They are all present in the Inconscient out of which they evolve. "Only what is involved can evolve." (EPY 560) Yet, the establishment of one level, or step, in the evolution requires a total concentration of the manifesting power on that step, a limitation to its potencies, otherwise nothing could be stable, fixed, moulded. This means that each level of evolution – matter, life, mind – has a 'ceiling' which it cannot pierce; for a higher level to be integrated upon Earth a response from the corresponding typal level is necessary. Evolution is two-tiered: emergence form below, descent from above.

For life to appear on our planet, powers and being of the life worlds had to embody in its Matter. For the human being to walk on this Earth, the powers and beings of the mental worlds had to incarnate on it. "Man is a being from the mental worlds whose mentality works here involved, obscure and degraded in a physical brain." (EDH 158) The material embodiment restricts the free abilities of the typal worlds. "Immense ranges of powers, influences, phenomena descend covertly upon us from the Overmind and the higher mental and vital ranges, but of these only a part, a selection, as it were, or restricted number can stage and realize themselves in the order of the physical world." (LD 780)

"Man cannot by his own effort make himself more than man ... But, still, mental man can open to what is beyond him" and call for supramental powers "to work in him and to do what the mind cannot do." (EDH 170) Because the human being has awareness it can be conscious of this aspiration. In the less or not aware lower organisms, this aspiration might be called an 'urge' or 'need', as intuited by Lamarck. "In the subjective order, we find that what shapes itself to us as a life-intention, life-impulse, life-formation here, already exists in a larger, more subtle, more plastic range of possibilities, and these pre-existent forces and formations are pressing upon us to release themselves in the physical world also; but only a part succeeds in getting through and even that emerges partially in a form and circumstance more proper to the system of terrestrial law and sequence. This precipitation takes place, normally, without our knowledge." (LD 774)

If all this may seem rather abstract, its application to the present day search for factual   coherence in the theories of evolution becomes very concrete. Dogmatic Darwinian gradualism prevented paleontology from taking the fossil record at face value; the enormous gaps in it, the 'missing links', were *a priori* held to be provisional and were expected to be filled up when in course of time more transitory species would be found. Till some theorists, annoyed with the unreality of this view, decided to accept the logical conclusions suggested by the situation, namely that the gaps are real and will not go away. We remember the words of Ernst Mayr: "According to Darwinian theory, evolution is a populational phenomenon and should therefore be gradual and continuous. This should be true not only for microevolution [on the genetic scale] but also for macroevolution [on the scale of species and higher], and for the transition between the two. Alas, this seems to be in conflict with observation ... Discontinuities [i.e. gaps] are overwhelmingly frequent ... The discontinuities are even more striking in the fossil record. *New species appear in the fossil record suddenly, not connected with their ancestors by a series of intermediaries.* Indeed there are rather few cases of continuous series of gradually evolving species."[15]

In 1972 Stephen Jay Gould and Niles Eldredge came forward with their theory of punctuated equilibrium: species have appeared suddenly, fully formed, and have kept their form for the duration of their existence. (The most striking example of a sudden appearance of new phyla – a phylum is a distinct way of building an animal – is the so-called "Cambrian explosion", around 550 million years ago, when most of the still existing species originated in a relatively very short time, a 'geological moment'. The Cambrian fauna was unrelated to its predecessors on Earth, the pre-Cambrian or Ediacaran fauna. The sudden and decisive event of the astonishing Cambrian explosion shocked the world of knowledge, and even made the cover of *Time* magazine in 1995.

Still more recent research ties the evolution of life, of its complexity and increasing consciousness, to the dramatic catastrophes our Earth has been subject to.[16] Life survived periods of total glaciation, when the planet looked like a snowball, and of infernal increases in heath after gigantic volcanic eruptions of its inner fire and impacts from meteorites. It also survived a quasi continuous shifting of the land masses, now called continents. Through all that it developed from microscopic molecules to (eukaryotic) cells with a nucleus, to conglomerations of cells, structured organisms, the countless species which are extinct, and those that still exist.

Each time nature has produced something unexpected, spectacular, sudden, 'magical', for which science has many hypotheses but no valid explanation. In Sri Aurobindo's view of a two-tiered evolution, the changes are not consequences effectuated only in matter: they are

---

15. Ernst Mayr: *What Evolution Is*, p. 208 (italics added).
16. See the recent series of Japanese documentaries realized with international cooperation: *Planet Earth* – not to be confused with David Attenborough's similarly named but older documentaries made for the BBC.

prepared on other levels of existence, vital and mental, and embodied in matter when life on Earth is ready for them. To our reason and logic these changes may appear incomprehensible, bizarre, and for the most part superfluous, if not bungled. But all are products of nature's "laboratory", and Nature, on Earth the material manifestation and gradual incarnation of the One, "makes no mistakes." Our science is the effort of our limited mentality, to which the ground of things, as well as their beginning and end, is not yet comprehensible.

"Thus the whole view of evolution begins to change," concluded Sri Aurobindo. "Instead of a mechanical, gradual, rigid evolution out of indeterminate Matter by Nature-Force we move towards the perception of a conscious, supple, flexible, intensely surprising and constantly dramatic evolution by a superconscient Knowledge which reveals things in Matter, Life and Mind out of the unfathomable Inconscient from which they rise." (EPY 174) What the theoretical mind of scientific materialism still holds to be a play of 'chance', of inexplicable coincidences, acquires meaning. "But what is Chance, after all? It is only a word, a notion formed by our consciousness to account for things of which we have no true knowledge – and it does not account for them." (EDH 299)

Homo sapiens

"The animal is a living laboratory in which Nature, it is said, worked out man," wrote Sri Aurobindo. (LD 4) In spite of the opinion of some learned men who think that they could have done better, the human product from nature's workshop does not seem too bad. "An Upanishad declares that the Self or Spirit after deciding on life-creation first formed animal kinds like the cow and the horse, but the gods – who are in the thought of the Upanishads powers of Consciousness and powers of Nature – found them to be insufficient vehicles, and the Spirit finally created the form of man which the gods saw to be excellently made and sufficient and they entered into it for their cosmic functions. This is a clear parable of the creation of more and more developed forms till one was found that was capable of housing consciousness." (LD 837)

Nature's "mental being", like the other fixed forms or species it has worked out, cannot by itself evolve into a higher embodiment of consciousness. "All the facts show that a type [i.e. a species] can vary within its own specification of nature, but there is nothing to show that it can go beyond it." While 'Darwinism' assumes smooth gradual transitions between the species, the fossil record, as we have seen, suggests otherwise – and so did Sri Aurobindo. Emergence from below and descent from above seems to be the explanation of all distinct forms of speciation. "It has not yet been really established that ape-kind developed into man; for it would rather seem that a type resembling the ape, but always characteristic of itself and not of apehood, developed within its own tendencies of nature and became what we know of man, the present human being." (LD 829)

"The type resembling the ape but always characteristic of itself" seems to have appeared 6 million years ago, according to the latest, sometimes very contradictory, interpretations of the fossil finds. The various types of "hominids" which then followed are called "australopithecines;" they walked upright, and the size of their brains increased steadily. "There were two major spurts of brain enlargement, one between 2 and 1.5 million years ago, which seems to be related to the appearance of *Homo habilis*, and a less pronounced one between 500,000 and 200,000 years ago."[17] "The earliest known [sub-species of] *Homo* first appear in the fossil record about 2.5 million years ago, around the age of the oldest known stone tools. They are different from other hominids in some striking ways: the have opposable thumbs and big brains. ... Modern humans evolved in Africa between 200,000 and 100,000 years ago."[18] At least two other human species coexisted for some time with *Homo sapiens*, the Neandertals, and the little people whose bones were lately discovered on an island in the Indonesian archipelago, *Homo floresiensis*. This brief outline of the evolution leading to our species as it is today may illustrate "the appearance in animal being of a type similar in some respects to the ape-kind but already from the beginning endowed with the elements of humanity." (LD 842)

"It is quite conceivable," wrote Sri Aurobindo, "that such an evolution from below and such a descent from above cooperated in the appearance of humanity in earth-nature. The secret psychical entity already there in the animal might have itself called down the mental being, the Mind-Purusha, into the realm of living Matter in order to take up the vital-mental energy already at work and lift it into a higher mentality. But this would still be a process of evolution, the higher plane only intervening to assist the appearance and enlargement of its own principle in terrestrial Nature." (LD 840)

---

17. Steven Mithen: *The Prehistory of the Mind*, p. 8.
18. Carl Zimmer: *Evolution*, p. 321, 360.

"Man is a type among many types so constructed, one pattern among the multitude of patterns in the manifestation in Matter. He is the most complex that has been created, the richest in content of consciousness and the curious ingeniousness of his building; he is the head of the earthly creation, but he does not exceed it. Even as others, so he too has his own native law, limits, special kind of existence; within those limits he can extend and develop, but he cannot go outside them. ... He is what he always was in the early beginnings of civilization; he continues to manifest the same capacities, the same qualities and defects, the same efforts, blunders, achievements, frustrations. If progress there has been, it is in a circle, at most perhaps in a widening circle. ... Nothing warrants the idea that he will ever hew his way out of the half-knowledge half-ignorance which is the stamp of his kind, or, even if he develops a higher knowledge, that he can break out of the utmost boundary of the mental circle. " (LD 831-32)

"We are in respect to our possible higher evolution much in the position of the original Ape of the Darwinian theory. It would have been impossible for that Ape leading his instinctive arboreal life in primeval forests to conceive that there would be one day an animal on the earth who would use a new faculty called reason upon the materials of his inner and outer existence, who would dominate by that power his instincts and habits, change the circumstances of his physical life, build for himself houses of stone, manipulate Nature's forces, sail the seas, ride the air, develop codes of conduct, evolve conscious methods for his mental and spiritual development. And if such a conception would have been possible for the Ape-mind, it would still have been difficult for him that by any progress of Nature or long effort of Will and tendency he himself could develop into that animal." (LD 55)

We have noted repeatedly with how much malice positivist biologists have brought the human down from the supreme position among creatures he was supposed to occupy before the life sciences converted to evolution. "We are merely neotenous apes that happen to be slightly cleverer than our cousins." (V.S. Ramachandran[19]) "*Homo sapiens* is a mammal and a primate, a member of the Class Mammalia and the order Primates that includes the monkeys, apes, and their kin. The forces of evolution that operate on other kinds of organisms have shaped humanity just as inexorably, and they continue to do so today." (Burton

---

19. John Brockman (ed.): *What is Your Dangerous Idea?* p. 23.

Guttman[20]) "Since Darwin, it is no longer useful to ask: 'Why has a particular species been created?' It is not scientifically productive to assume that the huge panoply of millions of species exists with regard to and somehow because of human beings. Similarly, it is no longer useful to suppose that we, as individuals, are the center of the universe, either." (David Barash[21])

Yet the human, evolved from the animal, is more than the animal: he is a mental being to a degree which essentially differs from that of the nascent mentality in the animal. It is the powers of the mind-world which have caused his body to walk upright, his brain to expand and his throat to produce language. "Man, because he has acquired reason and still more because he has indulged his power of imagination and intuition, is able to conceive an existence higher than his own, and even to envisage his personal elevation beyond his present state into that existence. His idea of the supreme state is an absolute of all that is positive to his own concepts and desirable to his own instinctive aspiration – Knowledge without its negative shadow of error, Bliss without its negation in experience of suffering, Power without its constant denial by incapacity, purity and plenitude of being without the opposing sense of defect and limitation. It is so that he conceives his gods; it is so that he constructs his heavens. ... His dream of God and Heaven is really a dream of his own perfection." (LD 55-6)

> He hangs between; in doubt to act or rest;
> In doubt to deem himself a god, or beast ...
> Created half to rise, and half to fall;
> Great lord of all things, yet a prey of all;
> Sole judge of truth, in endless error hurl'd;
> The glory, jest, and riddle of the world!
>
> (Alexander Pope, 1688-1744)

The human, seen not only as his material body but in the total composition of his being, is a "microcosm," thus called of old by people who seemed to know. The evolutionary layers which have built up the human are integrated into his constitution. The centres through which they support him, and through which he can enter into contact with them, are called *chakras* in India. They are located in the ethereal

---

20. Burton Guttman: *Evolution*, p. 141.
21. Alan Grafen and Mark Ridley (ed.): *Richard Dawkins*, p. 259.

substance of the spine. At its bottom is the centre of matter, with above it the lower vital or sex centre. Behind the navel is the centre of the middle vital (anger, fear), and behind the heart that of the higher vital, of the noble feelings. In the throat is the centre of mental expression, and between the eyes that of mental concentration (the third eye). And just above the head is the "lotus with a thousand leaves" which opens to the levels which earthly beings normally do not yet reach .

What is more, in the human too there is the presence of the One without which nothing can exist. "This spark of Divinity is there in all terrestrial living beings from the earth's highest to its lowest creatures." (LY 281) True spirituality is not a matter of superstition or mental negation of facts the mind cannot grasp: it is a matter of experience, prepared by longtime patient attention and accurate perception. Science has its value within the limits it has imposed on itself. Of another kind is the value of the experiences and formulations of the countless seekers after truth in their life-long exploration of domains which differ from those acknowledged by the Western mind.. The constitution of the human being is complex; this complexity is upheld by the centre of his being, "the divine spark" in him, his soul or psychic being, which is the cause and reason of his existence. Again, the truth of this statement can only be ascertained by direct personal experience.

This is a third view, other than scientific materialism and the Christian doctrine. It holds that evolution is real, because it is a movement of and within the Real or One, and that the human being is its highest evolutionary attainment, though not the final one. "We believe in the constant progression of humanity and we hold that that progression is the working out of a Thought in Life which sometimes manifests itself on the surface and sometimes sinks below and works behind the mask of external forces and interests. When there is this lapse below the surface, humanity has its periods of apparent retrogression or tardy evolution, its long hours of darkness or twilight during which the secret Thought behind works out one of its phases by the pressure mainly of economic, political and personal interests ignorant of any deeper aim within. When the Thought returns to the surface, humanity has its periods of light and of rapid efflorescence, its dawns and splendid springtides; and according to the depth, vitality, truth and self-effective energy of the form of Thought that emerges is the importance of the stride forward that it makes during these Hours of the Gods in our terrestrial manifestation." (EPY 140)

"The animal is a living laboratory in which Nature has, it is said, worked out man. Man himself may well be a thinking and living laboratory in whom and with whose conscious cooperation she wills to work out the superman, the god." (LD 4)

"The impersonal truth of things can be rendered into the abstract formulas of the pure reason, but there is another side of truth which belongs to the spiritual or mystic vision and without that inner vision of realities the abstract formulation of them is insufficiently alive, incomplete. The mystery of things is the true truth of things; the intellectual presentation is only truth in representation, in abstract symbols, as if in a cubist art of thought-speech, in geometric figure. It is necessary in a philosophic inquiry to confine oneself mostly to this intellectual presentation, but it is as well to remember that this is only the abstraction of the Truth and to seize it completely or express it completely there is needed a concrete experience and a more living and full-bodied language." (LD 357)

"A theory of spiritual evolution is not identical with a scientific theory of form-evolution and physical life-evolution; it must stand on its own inherent justification; it may accept the scientific account of physical evolution as a support or element, but the support is not indispensable. The scientific theory is concerned only with the outward and visible machinery and process, with the detail of Nature's execution, with the physical development of things in Matter and the law of development of Life and Mind in Matter; its account of the process may have to be considerably changed or may be dropped altogether in the light of new discovery; but that will not affect the self-evident fact of a spiritual evolution, an evolution of Consciousness, as progression of the soul's manifestation in material existence." (LD 836)

"Earth-life is not a lapse into the mire of something undivine, vain and miserable, offered by some Power to itself as a spectacle or to the embodied soul as a thing to be suffered and then cast away from it; it is the scene of the evolutionary unfolding of the being which moves towards the revelation of a supreme spiritual light and power and joy and oneness, but includes in it also the manifold diversity of the self-achieving spirit. There is an all-seeing purpose in the terrestrial creation; a divine plan is working itself out through its contradictions and perplexities which are the sign of the many-sided achievement towards which are being led the soul's growth and the endeavour of Nature." (LD 680)

Bibliography

Allègre, Claude: *Dieu face à la science*, Librairie Arthème Fayard, 1997

Amzallag, Gérard Nissim: *La Raison malmenée*, CNRS Editions, 2002

 — *L'Homme végétal*, Albin Michel, 2003

Arsac, Jacques: *La science et le sens de la vie*, Fayard, 1993

Baxter, Stephen: *Revolutions in the Earth*, Phoenix, 2003

Bassler, Moritz, and others (ed.): *Mystique, mysticisme et modernité*, PUS, 1998

Barash, David: *Sociobiology – The Whisperings Within*, Fontana/Collins, 1981

Bauer, Henry H.: *Scientific Literacy and the Myth of the Scientific Method*, University of Illinois Press, 1992

Behe, Michael J.: *Darwin's Black Box*, Free Press, 2006

 — *The Edge of Evolution*, Free Press, 2007

 — and others: *Science and Evidence for Design in the Universe*, Ignatius Press, 2000

Bensaude-Vincent, Bernadette, and others (ed.): *Des savants face à l'occulte*, Éditions la Découverte, 2002

Berlinski, David: *The Devil's Delusion*, Crown Forum, 2008

Black, Edwin: *War against the Weak*, Dialog Press, 2003

Blackmore, Susan: *The Meme Machine*, Oxford University Press, 1999

Bohm, David, and others: *Science, Order, and Creativity*, Bantam Books, 1987

Brockman, John (ed.): *The Next Fifty Years*, Phoenix, 2002

 — *The Third Culture*, A Touchstone Book, 1996

 — *What is Your Dangerous Idea?* Pocket Books, 2006

 — *What Are You Optimistic About?* Simon & Schuster, 2007

Brown, Andrew: *The Darwin Wars*, Touchstone, 1999

Capra, Fritjof: *Uncommon Wisdom*, Flamingo, 1983

Carey, John (ed.): *The Faber Book of Science*, Faber and Faber, 1995

Chadwick, Owen: *The Secularization of the European Mind in the 19th Century*, Cambridge University Press, 1975

Chown, Marcus: *Quantum Theory Cannot Hurt You*, Faber and Faber, 2007

Cobb, Matthew: *The Egg & Sperm Race*, Pocket Books, 2006

Collins, Francis: *The Language of God*, Pocket Books, 2007

Crombie, A.C.: *Medieval and Early Modern Science*, Doubleday Anchor Books, 1959

Darwin, Charles: *The Origin of Species*, Goyal Publishers, 1992

Davis, Paul: *God and the New Physics*, Penguin Books, 1983

 — *The Mind of God*, Penguin Books, 1993

 — *The Origin of Life*, Penguin Books, 1999

Dawkins, Richard: *A Devil's Chaplain*, Phoenix, 2004

 — *Climbing Mount Improbable*, Penguin Books, 2006

 — *River out of Eden*, Phoenix, 1999

 — *The Blind Watchmaker*, Penguin Books, 1988

   — *The God Delusion*, Bantam Press, 2006
   — *The Greatest Show on Earth*, Free Press, 2009
   — *The Selfish Gene*, Oxford University Press (Indian edition), 1989
   — *Unweaving the Rainbow*, Penguin Books, 1998
Delumeau, Jean (ed.): *Le savant et la foi*, Flammarion, 1989
Dembski, William A. (ed.): *Uncommon Dissent*, ISI Books, 2004
Dennett, Daniel C.: *Breaking the Spell*, Allen Lane, 2006
   — *Darwin's Dangerous Idea*, Penguin Books, 1996
Denton, Michael: *Evolution – A Theory in Crisis*, Adler & Adler, 1985
Dunbar, Robin: *The Trouble with Science*, Faber and Faber, 1996
Dyson, Freeman: *Infinite in All Directions*, Pelican Books, 1989
Ferguson, Kitty: *The Fire in the Equations*, Bantam Books, 1994
Foster, John Bellamy, and others: *Critique of Intelligent Design*, Monthly Review Press, 2008
Fox Keller, Evelyn: *The Century of the Gene*, Harvard University Press, 2000
Fuller, Steve: *Kuhn vs Popper*, Icon Books, 2003
Garreau, Joel: *Radical Revolution*, Doubleday, 2005
Gingerich, Owen: *God's Universe*, The Belknap Press (HUP), 2006
Gjertsen, Derek: *Science and Philosophy – Past and Present*, Penguin Books, 1989
Goodrick-Clarke, Nicholas: *The Occult Roots of Nazism*, New York University Press, 1994
Gould, Stephen Jay: *Punctuated Equilibrium*, The Belknap Press (HUP), 2007
   — *Rocks of Ages*, Ballantine Books, 1999
Grafen, Alan, and others (ed.): *Richard Dawkins*, Oxford University Press, 2006
Greenspan, Louis, and others: *Russell on Religion*, Routledge, 1999
Gribbin, John: *The Fellowship*, Penguin Books, 2006
Gribbin, Mary and John: *Being Human*, Phoenix, 1995
Guittès, Emmy: *Le passage de la matière à la vie*, La Baconnière, 1966
Guttman, Barton S.: *Evolution*, Oneworld Publications, 2007
Harris, Sam: *The End of Faith*, W.W. Norton & Company, 2004
Hellman, Hal: *Great Feuds in Science*, John Wiley & Sons, 1998
Hitchens, Christopher: *God is not Great*, Twelve, 2007
Hitching, Francis: *The Neck of the Giraffe*, Pan Books, 1982
Horgan, John: *Rational Mysticism*, Houghton Mifflin Company, 2003
   — *The End of Science*, Broadway Books, 1997
   — *The Undiscovered Mind*, The Free Press, 1999
Johnson, Phillip E.: *The Wedge of Truth*, InterVarsity Press, 2000
King, Francis, and others: *The Rebirth of Magic*, Corgi Books, 1982
Koestler, Arthur: *The Ghost in the Machine*, Arkana, 1989
Kuhn, Thomas S.: *The Structure of Scientific Revolutions*, The University of Chicago Press, 1996
Lafon, Claude: *Idées reçues en biologie*, Éllipses, 2004
Larson, Edward J.: *Evolution*, The Modern Library, 2004
Laurent, Goulven: *La naissance du transformisme*, Vuibert/Adapt, 2001

Lennox, John: *Hat die Wissenschaft Gott begraben?* R. Brockhaus Verlag, 2003
Lewens, Tim: *Darwin*, Routledge, 2007
Lewontin, R.C.: *The Doctrine of DNA*, Penguin Books, 1992
— *The Triple Helix*, Harvard University Press, 2000
Lindley, David: *Uncertainty*, Anchor Books, 2008
Lovejoy, Arthur: *The Great Chain of Being*, Harvard University Press, 1964
Mayr, Ernst: *What Evolution Is*, Phoenix, 2002
McCalman, Iain: *Darwin's Armada*, Simon & Schuster, 2009
McGrath, Alister: *Dawkins' God*, Blackwell Publishing, 2005
— *The Dawkins Delusion*, SPCK, 2007
Midgley, Mary: *Evolution as a Religion*, Routledge, 2002
Mithen, Steven: *The Prehistory of the Mind*, Phoenix, 2003
Monod, Jaques: *Le hasard et la nécessité*, Éditions du Seuil, 1970
Morange, Michel: *Les secrets du vivant*, Éditions la Découverte, 2005
Morris, Desmond: *The Human Zoo*, Dell Publishing Co., 1969
— *The Naked Ape*, Corgi Books, 1967
Noble, Denis: *The Music of Life*, Oxford University Press, 2006
O'Leary, Denyse: *By Design or by Chance*, Augsburg Books, 2004
Ortoli, Sven, and others: *Le cantique des quantiques*, La Découverte/Poche, 2004
Pauwels, Louis, and others: *Histoires magiques de l'histoire de France*, J'ai lu, 1980
Pichot, André: *Histoire de de la notion de géne*, Flammarion, 1999
— *Histoire de la notion de vie*, Gallimard, 1993
— *La société pure de Darwin à Hitler*, Flammarion, 2001
Porter, Roy: *Flesh in the Age of Reason*, Penguin Books, 2004
Prigogine, Ilya: *La Fin des certitudes*, 1998
— *Order out of Chaos*, Bantam Books, 1984
Rose, Hilary and Steven (ed.): *Alas, Poor Darwin*, Vintage, 2001
Rothman, Tony, and others: *Doubt and Certainty*, Basic Books, 1998
Ruse, Michael: *Darwinism and its Discontents*, Cambridge University Press, 2006
Sagan, Carl: *The Demon-Haunted World*, Headline, 1996
— *The Varieties of Scientific Experience*, The Penguin Press, 2006
Schumacher, E.F.: *A Guide for the Perplexed*, Abacus, 1977
Segerstråle, Ullica: *Defenders of the Truth*, Oxford University Press, 2000
Shanks, Niall: *God, the Devil, and Darwin*, Oxford University Press, 2004
Shapiro, Robert: *Origins*, Bantam Books, 1987
Shermer, Michael: *In Darwin's Shadow*, Oxford University Press, 2002
Simmons, Geoffrey: *What Darwin Didn't Know*, Harvest House Publishers, 2004
Singh, Simon: *Big Bang*, Harper Perennial, 2005
Smolin, Lee: *The Trouble with Physics*, Allen Lane, 2006
Sonigo, Pierre, and others: *L'évolution*, EDP Sciences, 2003
Sri Aurobindo: *Autobiographical Notes*, Complete Works, 2006
— *Essays Divine and Human*, Complete Works, 1997
— *Essays in Philosophy and Yoga*, Complete Works, 1999
— *Letters on Yoga*, Birth Centenary Library, 1970

— *Savitri*, 4th edition, 1993
— *The Human Cycle*, 3rd edition, 1998
— *The Life Divine*, Birth Centenary Library, 1972
— *The Synthesis of Yoga*, 6th edition 2nd impression, 1980.
Stannard, Russell: *Science and Wonders*, Faber and Faber, 1996
Staune, Jean, and others: *L'Homme face à la science*, Criterion, 1992
Stenger, Victor J.: *God – The Failed Hypothesis*, Prometheus Books, 2007
Stengers, Isabelle, and others: *100 mots pour commencer à penser les sciences*, Le Seuil, 2003
Sterelny, Kim: *Dawkins vs. Gould*, Icon Books, 2007
Susskind, Leonard: *The Cosmic Landscape*, Back Bay Books, 2006
Swyngedauw, Jean: *Á l'origine de la vie le hasard?* Éditions de l'O.E.I.L., 1990
Talbot, Michael: *Beyond the Quantum*, Bantam Books, 1988
Tresmontant, Claude: *L'histoire de l'univers et le sens de la création*, Tempus, 2006
Van Vrekhem, Georges: *Hitler and his God*, Rupa & Co., 2006
Weinberg, Steven: *Dreams of a Final Theory*, Vintage, 1993
— *The First Three Minutes*, Bantam Books, 1984
Wilber, Ken: *The Essential Ken Wilber*, Shambala, 1998
— *The Eye of Spirit*, Shambala, 1998
Wilber, Ken (ed.): *Quantum Questions*, New Science Library, 1984
Wilson, A.N.: *God's Funeral*, Abacus, 1999
Wilson, Edward O.: *Naturalist*, Penguin Books, 1996
Witham, Larry: *By Design – Science of God*, Unistar, 2003
— *The Measure of God*, HarperSanFrancisco, 2006
Woit, Peter: *Not Even Wrong*, Vintage Books, 2007
Wolpert, Louis: *The Unnatural Nature of Science*, Faber and Faber, 1993
Wright, Robert: *The Moral Animal*, Vintage Books, 1995
Ziman, John: *Reliable Knowledge*, Cambridge University Press, 1992
Zimmer, Carl: *Evolution*, Arrow Books, 2005
— *Soul Made Flesh*, Free Press, 2005.

Biographical Note

Georges Van Vrekhem is a Flemish speaking Belgian who writes in English. He became well-known in his country as a journalist, poet and playwright. For some time he was the artistic manager of a professional theatre company. He gave numerous talks and presentations in America, Europe and India.

He got first acquainted with the works of Sri Aurobindo and the Mother in 1964. In 1970 he joined the Sri Aurobindo Ashram in Pondicherry (now Puducherry), and in 1978 he became a member of Auroville, where he is still living and writing.

He wrote:

Beyond Man, the Life and Work of Sri Aurobindo and The Mother (1997)

The Mother, The Story of Her Life (2000)

Overman, the Intermediary between the Human and the Supramental Being (2000)

Patterns of the Present, in the Light of Sri Aurobindo and The Mother (2001)

The Mother: The Divine Shakti (2003)

Hitler and his God – The Background to the Nazi Phenomenon (2006).

Books by him are translated into Dutch, French, German, Italian, Russian, and Spanish. He was awarded the Sri Aurobindo Puraskar for 2006 by the Government of West Bengal.

Index

2050 (prognoses) 87, 193, 202
Abominable Snowman 234
Abrahamic religions 212, 231
Absolute Truth 185
Abstract thinking 88
Acquired characteristics 18, 19, 42, 59, 60, 61, 71, 102, 110
Adam 24, 91
Adams, Douglas 173
Adaptationists 254
Aegyptopithecus 83
Aesthetic appreciation 88
African-American 136
African-Americans 135
Age of Reason 45
Age of the Earth 39, 102, 113
Agnosticism 30, 45, 120, 146
Agnostic materialism 35
Agnostic materialist 216
Aitareya Upanishad 10
Alas Poor Darwin 16, 61, 144, 159, 198
Alchemy 78, 204, 223, 261
Alcock, John 139
Alice in Wonderland 21
Allègre, Claude 9, 13, 36, 37, 52, 272
Alvarez, Walter 166
Amazon 63, 65, 66, 67, 71
Amazonia 29
American-Indians 136
American mastodon 56
American Society for Psychical Research 82
Amoeba 54
Amzallag, Gerard 118, 184, 186, 255
Anaxagoras of Miletus 10
Anglican clergy 14
Anglo-Saxon Clubs 136
Animal Behaviour: An Evolutionary Approach 139
Annals and Magazine of Natural History 68

Anthropic principle 167, 232, 237, 238, 239, 240, 257
Anti-Lamarckism 61
Anti-religious evolutionists 42
Ants 140
Apollo space programme 136
Apostle of Unbelief 32
Arch-Darwinist 36
Arianism 211
Aristotelian worldview 183
Aristotle 50, 59, 62, 90, 91, 92, 182, 183, 186
Arsac, Jacques 229
Astronomy 230
Atheism 20, 21, 42, 119, 142, 146, 158, 159, 208, 209
Atkins, Peter 143, 170, 215, 218
Atran, Scott 43, 83, 119
Attenborough, David 245, 276
Aunger, Robert 156
Aurobindo, Sri 259-298
Auschwitz 137
Australopithecines 245
Australopithecus 40
Automata 53, 176
Avatars 10, 103, 261, 262
Avery, Oswald 117
Bacon's inductive method 80
Baltimore, David 203
Balzac, Honoré de 78
Barash, David 123, 124, 139, 140, 143, 145, 280
Bates, Henry 65, 67
Baudelaire, Charles 78
Bauer, Henry 167, 179, 185, 204
Becquerel, Henri 79
Behe, Michael 11, 35, 38, 74, 86, 122, 173, 190, 196, 237, 247, 248, 249, 250, 251, 252, 254, 255, 256
Belief in God 21, 120, 158, 225
Bergson, Henri 78, 187

Berlinski, David 37, 41, 42, 71, 72, 120, 221, 229, 234
Bible 14, 22, 24, 30, 102, 104, 159, 178, 209, 210, 241, 244, 245, 250
  interpretations of 24
  myths of 104
Biblical Flood 56
Biblical story 244
Big Bang 24, 178, 219, 238, 270, 271, 272
Bigfoot 234
Biochemistry 258
Biogeography 67, 68
Biological evolution 126
Biologie de Dieu 216
Biophores 110, 111, 120, 192, 193
Black, Edwin 44, 132, 134, 135, 136
Blackmore, Susan 149, 154, 155, 156
Black Order of the Death's Head 138
Blacks 135
Blind replicators 148
Blind Watchmaker, The 145, 146, 147, 157, 171, 189, 251, 253, 255
Blood of a Nation 135
Bloom, Floyd E. 87
Boar 103, 261
Bohm, David 175, 253, 257
Bohrian concept 266
Bohr, Niels 88, 221, 222, 241
Bolingbroke, Viscount 94
Book of British Birds 226
Boston Museum 40
Botanical Garden 47, 51
Bounty 15
Brachycephalic 135
Brahman 89, 212, 236, 260, 261, 262, 264, 269
Brain 44, 45, 57, 71, 72, 83, 87, 100, 149, 154, 155, 158, 190, 195, 216, 221, 252, 254, 263, 275, 278, 280
Breaking the Spell 42, 43, 159, 208, 215
Brenner, Sydney 202
Brief History of Time 233
British Society for Psychical Research 82
Brno 107
Brown, Andrew 169, 170
Brown, Jerram 139
Bruno, Giordano 21, 25, 211
Buddha 104, 235, 261, 262, 270

Buddhism 89, 222
Buffon, Georges Louis Leclerc, Count of 16, 24, 39, 47, 50, 52, 54, 95, 246
Cambrian 102, 166, 168, 230, 276, 289
Cambrian explosion 102, 166, 230, 276
Campaign to Create a Master Race 44, 132
Cape Town 26
Capivi, balsam of 66
Capra, Fritjof 97, 98
Carey, John 24, 43
Carnegie 135, 137
Carter, Brandon 239
Catastrophism 56
Catholic Church 160, 174, 183, 210, 213
Catholic faith 210
Catholicism 20, 210
Catholics 120, 134, 225
Caucasian 125, 135, 136
Causes of Evolution 114
Cell nucleus 109
Chadwick, Owen 31
Chain of Being 16, 45, 54, 74, 75, 86, 88, 89, 91, 92, 95, 97, 256, 267, 269
Chakras 280
Chambers, Robert 26, 65, 67, 131, 246
Chance and necessity 142
Chardin, Pierre Teilhard de 12, 120, 145, 245
Chimpanzees 39
Chosen People 212
Chown, Marcus 180
Christ 28, 209, 245, 248, 262
Christian Churches 77
Christian creationism 249
Christian duality 91
Christian fundamentalism 160
Christian God 21, 64, 160, 226, 257, 260
Christianity 31, 209, 210, 212, 214, 226, 244, 259
Christian morality 211
Christian mysticism 235
Christian mystics 234
Christian theology 95, 244, 261
Chromosomes 107, 109, 114, 149, 192, 195, 202, 204
Church Fathers 104, 178, 209
Church of England 24

Church of Scientism 213, 224
Circumcision 110
Class Mammalia 82, 279
Clones 193
Cobb, Matthew 50, 186
Coelacanth 164
Cognition 216
Collins, Francis 225
Colonialism 136, 209
Columbus, Realdus 178
Communist Manifesto 26
Comoro Islands 164
Comte, Auguste 127
Copernican principle 123, 124, 167, 232, 266
Copernican worldview 123, 124, 167, 183, 232, 239, 266
Copernicus, Nicolaus 11, 30, 37, 45, 124, 143, 183, 185
Correns, Carl 111
Cosmic Magician 263
Cosmic soup 238
Cosmological Chain of Being 92
Creationism 12, 45, 52, 67, 95, 111, 113, 130, 160, 165, 211, 241, 243, 249, 250
Creator 9, 111, 131, 165, 198, 209, 233
Cretaceous 166
Crick, Francis 61, 117, 140, 180, 187, 192, 193, 199, 205
Critique of Intelligent Design 21, 43, 243
Crookes, William 79
Crusades 210
Curie, Pierre and Marie 79
Cuvier, Georges 16, 50, 51, 56, 95, 246
Dancing Wu Li Masters 221
D'Andréa, Patrizia 82
Dangerous idea 37, 83
Dart, Raymond 40
Darwin, Erasmus 17, 61, 95, 131, 246
Darwinian biology 12
Darwinism is dead 114
Darwin on Trial 242
Darwin's Armada 14, 32, 65
Darwin's Black Box 11, 38, 74, 86, 122, 196, 247, 249, 251, 254
Darwin's bulldog 32
Darwin's Finches 120
Darwin's religious struggle 45

Darwin's theory of sexual selection 71
Darwin Wars 161, 169, 170
Davenport, Charles 134, 136
Davies, Paul 9, 24, 39, 99, 100, 118, 226
Da Vinci, Leonardo 49, 186
Dawkins Delusion 158, 226
Day-age theory 244, 245
De Harmonia Mundi 223
De Humani Corporis Fabrica 186
Delhaye, Jean 225
De Magnete 178
Dembski, William 12, 120, 146, 160, 169, 190, 201, 248, 249, 250, 252, 256
Democritus 10
De Morgan, Augustus 79
Dennert, Eberhard 103
Dennett, Daniel 12, 34, 40, 41, 42, 43, 113, 155, 156, 159, 170, 209, 215, 216, 226, 253, 254
Denton, Michael 16, 23, 27, 30, 31, 35, 36, 38, 41, 195, 201, 242, 247
De Revolutionibus Orbium Celestium 30, 185
Descartes, René 21, 22, 50, 52, 87, 90, 174, 176, 180, 186, 211, 223, 240
Designer 196, 198
Deuteronomy 210
Devaluation of the human existence 167
Devonian 166
De Vries, Hugo 13, 41, 111, 112, 113, 120, 147, 192, 194
Diana, Princess of Wales 42
Dickens, Charles 30
Diderot, Dennis 16, 95
Dinosaurs 35, 164, 166, 167
Discourse on Method 176
DNA duplication 142, 193
Dobzhansky, Theodosius 101, 116, 120, 168
Dogmas 174, 204, 206, 228
Dogmatic religion 223
Dolomites 49
Dorst, Jean 121
Dose, Klaus 86
Double helix 13, 141, 146, 180, 193, 196, 202, 241
Dover, Gabriel 198
Doyle, Arthur Conan 78

Dreyfus Affair 133
Driesch, Hans 187
Dubois, Eugen 40
Dunbar, Robin 187, 218
Dwarf 103
Dyson, Freeman 233, 239, 240
Eagleton, Terry 226
Earth, age of 39, 102, 113
Eckhart, Meister 234
Economics 104, 105
Economic theory of Thomas Malthus 104
Eddington, Arthur 61, 221, 223
Egypt 51, 230, 268
Eibl-Eibesfeldt, Irenäus 139
Einstein, Albert 11, 38, 141, 158, 174, 180, 188, 219, 220, 221, 223, 241, 270
Einsteinian concept 183, 266
Eldredge, Niles 12, 40, 56, 102, 161, 162, 163, 164, 166, 276
Electrical charge of the proton 238
Electro-magnetic force 238
Elementary living particles 120
Elephant 54, 151
Elk 56
Embryonic development 9
Empiricism 20, 38, 216
Encyclopédie 16, 95
End of Faith 209, 210, 211, 235, 236
End of Physics 218
End of Science 167, 182, 190, 218
Enemy of the church 21
Engels 26
Enlightenment 16, 20, 31, 45, 49, 51, 81, 104, 209, 211, 223, 241
Epigenetics 60, 205, 207
Epiphenomenon 53, 57, 87, 156
Espinosa, Baruch de 212
Essay on Man 92
Essay on Population 29, 69
Essentialism 159, 160
Euclidean geometry 227
Eugenic movement 134
Eugenic propagandists 135
Eugenics 44, 127, 130, 131, 132, 133, 134, 135, 136, 137, 144, 169
Existence of God 21, 51, 228

Extrapolation backwards 25
Eye of Spirit 89
Fantastic materialism 220
Feynman, Richard 178, 222
Fine-tuned universe 238
First Three Minutes 142, 143
Fish 10, 92, 96, 103, 164, 261
Fisher, Ronald 116, 134
FitzRoy, Robert 14, 15
Fixism 52, 56
Flammarion, Camille 79
Fleming, Alexander 180
Forbidden Books 212
Forced sterilization 135
Ford, Henry 137
Four musketeers 32, 37, 104
France 20, 25, 49, 51, 52, 77, 87, 109, 127, 128, 133, 216
French Revolution 20, 51
Freud, Sigmund 61, 73, 78, 143, 243
Frisch, Karl von 140
Galen 50, 178, 186
Galileo Galilei 21, 25, 35, 76, 174, 175, 177, 183, 186, 210, 211, 212, 223, 234, 240, 241, 266
Galton, Francis 41, 109, 131, 132, 133
Gametes 109
Garreau, Joel 273, 284
Gauthama, Buddha 270
Gemmules 18, 59, 105, 108, 109, 110, 111, 112, 192
Genesis 244, 245, 250
Genetic determinism 170
Genetic essentialism 160
Genetics plus natural selection 116
Geology 23, 24, 25, 39, 67, 104, 105, 130, 244, 246
George III, King 17
Germain, Paul 225
Germ plasm theory 109
Giant ground sloth 56
Gifford Lectures 232
Gilbert, Walter 202, 203
Gilbert, William 178
Gilson, Etienne 209
Gingerich, Owen 149, 150, 159, 185, 224
Giraffe 58

God and the New Physics 100, 226
God as a Mathematician 233
God's existence 21
God's Universe 150, 159, 185, 224
Goodrick-Clarke, Nicholas 77
Gorillas 39, 40
Goslings 140
Göttingen 109
Graaf, Reinier de 187
Gradualism 25, 26, 37, 39, 56, 68, 71, 101, 102, 146, 161, 163, 165, 189, 246, 253, 257, 275
Gradual transformation 112
Grail of human genetics 203
Grand Unified Theory 175, 187, 233
Grant, Robert 14, 20
Gravitational force 238
Gray, Asa 29, 32, 101
Great Architect 233
Great Britons contest 42
Great Chain of Being 88, 89, 91, 92, 97
Greece 104, 230, 271
Greek philosophers 90
Greek reason 91
Gregor, Brother 107
Gribbin, John 83, 141, 142, 167, 179
Ground sloth 56
Guide for the Perplexed 97
Guittès, Emmy 270
Guttman, Burton 61, 82, 165, 280
Gypsies 133, 137
Hadewych 234
Haeckel, Ernst 37, 41, 101, 102, 109, 125, 127, 128, 137
Haldane, J.B.S. 114, 116, 117, 120
Hamilton, William 140, 145, 170
Hamlet 263
Handicapped 137
Harriman, E.H. 135
Harris, Sam 209, 210, 211, 235, 236
Harvey, William 178
Hasard et la nécessité 142, 146
Hawking, Stephen 124, 233
Hegel, Georg Wilhelm Friedrich 222
Heisenberg, Werner 88, 180, 221, 223
Hereditary Genius 131
Heredity, theory of 59, 108, 109, 112

Hermetism 261
Herschel, John 26
Higgs boson 219
Himmler, Heinrich 138
Hinduism 120, 212, 261
Histones 204
Hitchens, Christopher 159, 168, 209, 213, 230, 231
Hitchhikers Guide to the Galaxy 173
Hitler 44, 125, 127, 133, 136, 137, 138
Hobbes, Thomas 20, 21
Holism 175, 200
Holmes, Sherlock 78
Holocaust-deniers 136
Holy Book 211
Holy books 12, 212
Holy Grail 159, 201, 202, 203
Holy Spirit 211
Homo erectus 40
Hooker, Joseph 15, 29, 32, 33, 34, 69, 70, 80
Horgan, John 118, 167, 182, 190, 192, 194, 195, 216, 218, 228, 234, 235
Horne, Johannes van 187
Hugo, Victor 78
Human Ethology 139
Human genome 196, 202, 203, 225
Human Genome Project 225
Humanism 133, 138, 150
Humanitarian socialism 84, 170
Humboldt, Alexander von 20, 64, 65
Hume, David 20, 226
Hut, Piet 181
Hutton, James 23, 39, 50
Huxley, Aldous 116
Huxley, Arthur 234
Huxley, Julian 42, 116, 120, 133, 139
Huxley, Thomas 15, 26, 32, 116, 131, 163
Huysmans, Joris-Karl 78
Hybridization 107
Hypnotism 78
Hypothetico-deductive method 80
Icons of Evolution 122
Illuminati 78
Il Saggiatore 177
Immortality 159

Increasing complexity 16, 45, 49, 58, 95, 104, 195, 262

India 4, 40, 49, 83, 103, 196, 261, 262, 266, 268, 271, 280

Indian 10, 80, 135, 234, 284

Indonesia 66

Inductive method 80

Industrial revolution 26, 45, 81, 84, 134

Inheritance of characteristics 59, 110

Inquisition 174, 210, 223

Instinctive memory of evolution 103

Intelligent Design 21, 43, 73, 237, 240, 241, 242, 243, 244, 247, 248, 249, 250, 251, 255, 256, 257, 258

Intelligent Design Movement 248, 249, 257

Intelligent design theory 12, 237

Intelligent Power 193

Intracellular Pangenesis 111

Involution 263, 264, 274

Irish elk 56

Islam 10, 211, 212, 244

Israel 206, 213

Italy 128

Jacobins 51

Jaeger, Werner 212

Java Man 40

Jean-Baptiste, Patrick 216

Jericho 212

Jerusalem 38

Jesuits 91, 223

Jesus 28, 129, 209, 225, 262

Jews 133, 134, 136, 137, 138

Johnson, Phillip 242, 243, 250

Joshua 212

Judaism 212, 244

Judeo-Christian ethics 243, 249

Judeo-Christian God 160, 226, 257

Jung, Carl Gustav 61, 73

Junk DNA 196

Kabbalah 89

Kant, Immanuel 93, 222

Kapital, Das 30

Kardec, Allan 77

Kekulé, August 180

Keller, Fox 193, 194, 203, 206, 207

Kelvin, Lord 61, 102, 113

Kennedy, John President 136

Kepler, Johann 186, 223, 224, 241

Kingsolver, Barbara 12

Koestler, Arthur 98, 99, 120, 176

Koons, Robert 37, 41

Krishna 104, 261

Kuhn, Thomas 182, 183, 184, 185, 207, 213, 219, 224, 227, 260

Küng, Hans 212

Lafon, Claude 204

La fonction fait l'organe 58

Lagrange, Pierre 82

Lamarckism 12, 60, 61, 71, 102, 113, 131, 150, 199, 201, 206, 246

Lamarck, Jean-Baptiste de 14, 47, 50, 58, 110, 200, 206, 246

La Mettrie, Julien Offray de 44, 50, 53, 91

Language of God 225

Language of science 35, 177

Laplace, Pierre-Simon 21, 22, 23, 143

Lapouge, Vacher de 125, 127, 128

Larson, Edward 27, 30, 34, 101, 102, 103, 106, 107, 113, 115, 116, 117, 118, 120, 134, 193

Last Judgment 212, 231

Laurent, Goulven 18, 49, 50, 51, 52, 54, 55, 56, 57

Lavoisier, Antoine 51

Law of thermodynamics 126

Lazarus, Emma 135

Lebenswunder, Die 137

Leeuwenhoek, Antonie van 187, 240

Leibniz, Gottfried Wilhelm 56, 95

Leiden 50, 187

Leonardo da Vinci 49, 53, 186

Levels of awareness 89

Levels of being 77, 88, 89, 90, 91, 98, 266, 274

Lévi, Éliphas 78

Lewens, Tim 11, 15, 16, 21, 26, 27, 117, 123, 124, 125, 130, 258

Lewontin, Richard 106, 157, 169, 170, 175, 176, 177, 178, 181, 188, 190, 197, 198, 200, 202, 253, 255

Liberation theology 212

Life Force 198

Light, speed of 238

Lindbergh, Charles 137
Linley, David 218
Linnaeus, Carolus 16, 17, 18, 45, 48, 50, 54, 56, 93, 95, 130, 246
Linné, Karl von 16, 18
Lion 40
Livre des Esprits 77
Loch Ness Monster 234
Locke, John 20, 21
Lombroso, Cesare 79
Long Island 134, 137
Long-skulled 135
Lord of creation 16, 43, 95
Lorenz, Konrad 140
Lovejoy, Arthur 91, 92, 94, 95, 209
Lutherans 134
Lyell, Charles 23, 25, 28, 29, 32, 37, 39, 41, 50, 66, 67, 68, 69, 70, 73, 85, 101, 104, 189, 246
MacCalman, Iain 29
Macroevolution 164, 165, 171, 276
Macromutation 171
Macromutations 171
Macrophysics 219
Madagascar 136
Mahabharata 234
Maimonides 97
Malay Archipelago 29, 49, 63, 66, 68, 70, 75
Malaysia 29, 66
Mallarmé, Stéphane 78
Malthus, Thomas 26, 27, 29, 36, 41, 64, 67, 69, 104, 106
Man's Place in the Universe 83
Maoism 213
Margulis, Lynn 122
Martyrs 104
Marx, Karl 26, 30, 134, 243
Mass of the electron 238
Masterales 206
Master race 44, 132
Mastodon 56
Maternal effects 201
Mathematical reasoning 88
Mathematics 13, 35, 51, 82, 87, 113, 115, 134, 143, 177, 186, 187, 233, 260
Mayr, Ernst 34, 40, 103, 104, 116, 162, 164, 188, 189, 194, 195, 276

McCalman, Iain 14, 15, 32, 33, 65, 66, 68
McClintock, Barbara 195
McCormick, Robert 15
McCulloch, Oscar 135
McGinn, Bernard 235
McGrath, Alister 156, 157, 158, 225, 226
Mediocrity, Principle of 232
Mein Kampf 138
Meiosis 109
Meme 151, 153, 154, 155, 156, 170, 172, 246
Memes 152, 153, 154, 155, 156, 157, 197, 224
Memetic selection 155
Mendeleyev, Dmitri Mendeleev 180
Mendel, Gregor 13, 106, 107, 108, 110, 111, 113, 114, 140, 194
Mengele 137
Mesmerism 78
Metaphysical materialism 61, 141, 266
Microbiology 142, 193
Microevolution 163, 164, 276
Microphysics 219
Middle Ages 10, 31, 37, 78, 81, 92
Midgley, Mary 188, 198
Mimeme 153
Mind-body problem 72, 87, 93, 263
Mind of God 226
Mineral 87, 97, 103, 267
Miocene apes 83
Mitosis 109
Modern synthesis 115, 117, 118, 119, 120, 121
Moltke, Marshall von 129
Mongrelization 135
Monkey trial 211
Monks 107, 210, 223
Monod, Jacques 91, 142, 144, 146, 147, 152, 153, 161, 177, 242, 267
Montpellier 187
Morange, Michel 203
Morgan, Thomas 13, 114
Morris, Desmond 124, 139, 242
Multiverse 265, 271, 274
Muséum d'histoire naturelle 121
Mutation, theory of 112
Myology 186

Mysticism 80, 88, 111, 148, 187, 193, 206, 216, 221, 223, 228, 235, 259, 270
Nabi, Isidore 151
Nagarjuna 235
Naked Ape 124, 139, 242
Nanotechnology 193
Napoleon 22, 51
Natural theology 22, 35, 36, 42, 67, 111, 157, 212, 245
Nautilus pompilius 49
Nazi Germany 133, 137
Nazism 44, 77, 125, 127, 131, 133, 134, 137, 213
Negative eugenics 132
Negroes 65, 125, 135, 136
Nelkin, Dorothy 159
Neo-Darwinian theory 201
Neo-Darwinism 13, 45, 111, 115, 116, 118, 122, 145, 147, 162, 164, 168, 188, 189, 242, 243, 246
Neotenous ape 43
Neoteny 149
Nervous system 168, 181
Neuroscientists 216
Neurotheology 216
New Science 35
New synthesis 13, 115, 116, 117, 152
New Testament 226, 245
Newtonian physics 183
Newton, Isaac 11, 21, 22, 30, 38, 51, 56, 61, 65, 141, 174, 180, 183, 186, 211, 219, 223, 238, 240, 241
Newton's apple 180
Nietzsche, Friedrich 21, 78, 127, 131
Nobel Prize 79, 140, 142, 160, 192, 218, 219, 226
Noble, Denis 121, 153, 190, 191, 195, 197, 198, 199, 200, 201, 205
Non-constancy of species 104
Non-overlapping magisteria 217
Nordic superiority 134
Novicow, Jacques 129
Nuclear acids 193
Nuova scienza 35, 174, 178
Nuremberg Laws 137
Obama, Barack 136
Obscurantist holism 175, 200

Occultism 45, 77, 78, 80, 81, 82, 111, 174, 187, 261
Oceanic feeling 223
Old Testament 21, 160, 226, 245
O'Leary, Denyse 11, 12, 40, 42, 122, 158, 161, 162, 208, 248, 250, 251, 256, 257
OM 212
Omega Point theory 12, 145
Omnipresent Reality 236, 260
One-celled organisms 86
One-gene-one-effect doctrine 148
One-to-one theory 194
Orangutan 168
Orang-utans 39
Ordovician 166
Original Sin 91
Orthodox 120, 134, 168
Ortoli, Sven 219, 220
Overmind 275
Overruling Intelligence 72, 73, 74
Ovum 13
Owen, Richard 39, 56
Oxford Dictionary of Biology 11, 115, 196
Padmasambhava 235
Padua 50, 178, 187
Paleontological data 244
Paleontologists 162, 166, 173, 258
Paleontology 18, 49, 50, 56, 104, 114, 130, 162, 166, 170, 230, 244, 246, 275
Paley, William 22, 23, 42, 67, 146, 248
Pangenes 111, 112, 120, 192
Pangenesis 18, 59, 60, 105, 108, 109, 111
Papal infallibility 210
Paracelsus 50
Paris 24, 48, 82, 187
Pauli, Wolfgang Ernst 221, 223
Peking Man 40
Penicillin 180
Perennial philosophy 89, 90, 235
Perfection of the Absolute Being 92
Permian 166, 230, 295
Permian extinction 230
Physics envy 187
Pichot, André 13, 18, 19, 32, 36, 41, 42, 47, 48, 52, 55, 58, 59, 60, 61, 62, 101, 103, 105, 106, 107, 108, 112, 113, 115,

121, 125, 126, 127, 128, 129, 133, 142, 193, 194, 201, 205, 206, 246
Pikaia gracilens 168
Pinker, Stephen 42, 170
Planck's constant 238
Planet Earth 55, 86, 266, 272
Plato 87, 90, 91, 133, 222, 263
Platonic Absolute 212
Pleiotropy 195
Plessis, Anne 206
Plotinus 222
Poe, Edgar Allan 78
Poor laws 44
Pope, Alexander 92, 93, 280
Popper, Karl 182, 185, 213
Porco, Carolyn C. 215
Porete, Margarete 234
Positive eugenics 132
Prigogine, Ilya 218, 219
Principia Mathematica 30, 51, 223
Principle of continuity 92
Principle of plenitude 91, 92
Protein synthesis 142, 193
Protestant church 160
Protestantism 120, 170
Protozoa 54
Proust, Marcel 78
Psychical Research 82
Punctuated equilibrium 12, 13, 56, 117, 161, 162, 165, 171, 246, 276
Pythagoras 90, 222
Qualia 177
Quantum field theory 230
Quantum materialism 220
Quantum mechanics 88, 182, 188, 226, 227
Quantum syncretism 220
Queen Victoria 20, 78
Racial superiority 125
Racism 136
Racist nationalism 133
Railway fever 64
Rama 103, 261
Ramachandran, V.S. 279
Ramayana 234
Rational Mysticism 216, 228, 235
Rattlesnake 15

Ray, John 22
Rayleigh, Lord 79
Recapitulation theory 128
Redi, Francesco 186
Reductionism 121, 144, 174, 175, 207, 218, 245
Redwood 151
Reformation 31, 33
Reincarnation 83
Religion 24, 146, 159, 160, 188, 208, 209, 213, 216, 217, 218, 224, 226, 231, 235
Religious aggression 211
Religious obscurantism 81
Religious scientists 213, 244
Renaissance 31, 35, 53, 81, 104, 174, 223
Renan, Ernest 129
Rensch, Bernhard 188
Reproduction and alimentation 187
Republic 51, 133, 263
Richet, Charles 79
Rig Veda 86, 90
Rimbaud, Arthur 78
RNA 142, 192, 193, 195, 196, 197, 204, 205
Roberts, Royston 180
Robin 168
Robinet 95
Rockefeller, John D. 135, 137
Roman Catholics 225
Roman Empire 210
Roman Law 133
Roman Republic 133
Rome 212, 223, 230
Röntgen, Wilhelm 79
Rose, Steven 16, 61, 118, 144, 159, 170, 190, 198
Rosicrucianism 78, 261
Royal Society 157, 178, 187
Royer, Clémence 109, 127, 128
Rumi, Jalal ad-Din ar- 10
Ruse, Michael 9, 37, 43, 60
Russell, Bertrand 24, 91, 145, 146, 149, 150, 175, 190, 213, 215, 226, 231, 267
Russia 128
Russian communism 213
Sagan, Carl 88, 157, 209, 215, 226, 227, 230
Saltations 111

Saltation theories 102
Sarawak paper 68
Satan 211
Scale of Being 93, 95
Schillebeeckx, Edward 212
Schopenhauer, Arthur 222
Schrödinger, Erwin 221, 222, 223
Schumacher, E.F. 97
Schutzenberger, Marcel 43, 93, 252
Science fiction materialism 220
Scientific American 40, 87, 196, 216, 271
Scientific materialism 11, 42, 53, 63, 72, 74, 87, 88, 94, 98, 104, 121, 148, 149, 156, 176, 207, 214, 216, 232, 237, 240, 241, 242, 243, 245, 254, 263, 268, 277, 281
Scientific method 35, 51, 80, 120, 173, 174, 176, 177, 178, 179, 181, 187, 266
Scopes, John 211
Secrets du vivant 203
Segerstråle, Ullica 141, 143, 144, 151, 152, 178, 187
Selfish Gene 139, 146, 147, 151, 152, 153, 193, 196, 197, 200, 253
Selfishness 123, 148, 157, 224
Self-replication 198, 199
Sense organs 168
Serendipity 180
Shankara 235, 262
Shapiro, Arthur 169
Shapiro, Robert 55, 86, 122, 192, 193
Shark 151, 168
Shermer, Michael 17, 29, 30, 60, 63, 65, 67, 68, 70, 71, 72, 74, 75, 76, 79, 80, 83, 84, 88, 285, 297
Simmons, Geoffrey 252
Simonyi, Charles 157
Sloth 56
Small is Beautiful 97
Smith, Adam 27, 106
Smith, John Maynard 170
Smolin, Lee 190, 218, 219, 238
Snake 83, 253
Social biology 44
Social Darwinism 127, 129, 130, 131, 132, 139, 144, 169
Sociobiology 13, 120, 124, 139, 140, 141, 142, 143, 144, 145, 147, 150, 151, 152, 159, 169, 170, 187, 217, 242, 246

Socrates 10, 25
Soft sciences 110
Soup metaphor 153
South America 14, 15, 20
Speciation 34, 40, 165, 278
Spencer, Herbert 37, 44, 101, 109, 126, 128, 131
Spermatozoon 13, 107
Sperm cell 107
Spinoza, Baruch 212
Spiritism 75, 76, 77, 78, 79, 80
Spiritual alchemy 261
Spiritual Couplets 10
Spiritual experience 90, 235, 260, 262, 264
Spiritualist Congress 82
Spiritual realism 89
Steam locomotive 64
Steinhardt, Paul 271
Stengers, Isabelle 118, 179, 219
Stenger, Victor 208, 209, 216, 227, 228, 231, 234, 238
Steno, Niels 187
Sterelny, Kim 166, 199
Sterilization 135
Structure of Scientific Revolutions 182, 224
Struggle for life 123, 126, 128, 129
Struggle for survival 27
Sudarshan, George 240
Superconscient 277
Supermind 233, 234, 264, 265, 274
Supernatural 76, 80, 158, 160, 165, 189, 208, 224, 226, 234, 246, 256
Super race 134
Supremacist white 135
Supreme Being 51, 52, 157
Supreme Truth 211
Susskind, David 218, 239
Swammerdam, Jan 187
Swedenborg, Emanuel 78
Swyngedauw, Jean 118, 119, 120, 258
Symposium 263
Systema Naturae 16
Tables of the Roman Law 133
Talbot, Michael 188
Taoism 222

Tao of Physics 221
Taung Child 40
Teleological metaphor 72
Templeton Prize for Progress in Religion 226
Ten Commandments 210
Ternate island 67, 69
Terrorism 211
Tertiary 166
Thermodynamics 126
Thomson, William 102
Times of London 30
Tinbergen, Nikolaas 140, 145
Tortoise 103, 186, 261
Transformism 18, 56, 67, 95, 104
Transformist theory 54
Transmutationism 18, 56, 61, 67, 95, 104, 269
Transmutation of species 162
Transposon 195
Tree of life 35, 39, 48, 84, 124, 136
Tresmontant, Claude 149, 225, 245
Triassic 166
Tribes of Israel 213
Tschermiak, Erich von 111
Turok, Neil 271
Twelve Tables of the Roman Law 133
UNESCO 116, 133
Upanishads 10, 222, 260, 277
Uranus 26
Use creates the organ 58
Ussher, Archbishop James 24, 50
Valéry, Paul 78
Vedas 86, 90, 222, 262
Vedic knowledge 90
Vedic seers 90, 96
Verne, Jules 78
Vesalius, Andreas 50, 178, 186
Victorian Age 20, 126
Victorian England 30, 106
Victorian optimism 126
Vie de Jésus 129
Vietnam War 170
Vigier, Jean-Pierre 218

Vinge, Vernor 272
Vishnu 86, 90, 266
Vitalism 187, 240
Vitalists 144
Voltaire 51, 226
Wallace, Alfred 28, 29, 63, 65, 67, 68, 73, 78, 79, 80, 81, 83, 84, 88, 92, 131, 163, 246
War against the Weak 44
Wasps 140, 166
Watson, James 117, 140, 180, 193, 205
Wedge of Truth 243
Weinberg, Steven 142, 143, 160, 209, 218, 222, 223, 239
Weismann, August 13, 41, 73, 108, 109, 110, 111, 120, 147, 192, 199, 201
Weismann barrier 109, 192
Weismannism 108
Wells, Jonathan 118, 122
Westminster Abbey 226
White-skinned 35
Wilberforce, Samuel 33, 34
Wilson, A.N. 21, 30, 31, 37, 210
Wilson, Edward 120, 139, 140, 141, 143, 144, 145, 146, 147, 151, 159, 169, 170, 187, 193, 217, 242, 246
Wilson, Reverend 40
Witch hunts 210
Witham, Larry 87, 119, 143, 176, 177, 193, 209, 212, 232, 233, 238
Witte, Alfred 222
Woit, Peter 190, 238
Wolpert, Lewis 181
World War I 131
World War II 44, 241
X-rays 79
Yahweh 209, 210, 229
Yin-yang 222
Young, Edward 96
Zarathustra 21
Zeus 212
Ziman, John 179, 180, 184
Zimmer, Carl 21, 26, 34, 39, 65, 119, 232, 278
Zoonomia or the Laws of Organic Life 17

Printed in Great Britain
by Amazon